Leona's Legacy:
Memories of A Farmer's Daughter

By Kelly Simons

The Guest Cottage Inc.

Published by
The Guest Cottage • Woodruff, Wisconsin

ISBN 1-930596-17-0
Published by The Guest Cottage
PO Box 848 • Woodruff, WI 54568
800-333-8122 • www.theguestcottage.com

Please write or call to request a free catalog of other publications from The Guest Cottage.

Edited by The Word Forge • Ferndale, Pennsylvania

Cover and interior design by
Mary Shafer Creative Services • Ferndale, Pennsylvania

Printed in the United States of America

Library of Congress Cataloging-in-Publication Data

Simons, Kelly.
Leona's legacy : memories of a farmer's daughter / by Kelly Simons.
p. cm.
ISBN 1-930596-17-0 (pbk.)
1. Ahnen, Leona. 2. Farmers—Wisconsin—Biography.
3. Farm life—Wisconsin. 4. Depressions—1929—Wisconsin.
5. Wisconsin—Biography. I. Title.
CT275.A446 S56 2002
977.5'032'092—dc21

2002012805

Contents

Dedication

This book is dedicated to Lee Bennett,
and to the memory of her family.

Acknowledgments

Many people helped in bringing this book to fruition, either through direct involvement or through their love and support. I would like to thank them all, with special thanks to:

First, my dear friend, Lee Bennett, for her superb memory filled with childhood stories from seventy years ago. Thank you, Lee, for taking me to Boscobel to see the old homestead, for all the hours of conversation and laughter along the way, for your patience with me when I was asking a million and one questions, and for being such a good friend. Without you, there would be no book.

My publisher, Nancy Ravanelli, for taking a chance on an unknown writer like me.

My editor, Mary Shafer, for her hard work on my manuscript, her words of encouragement, and her guidance along the way.

My husband, Joe, and our children, Ben and Leah, for their love, encouragement, and patience during the past few years.

My parents, Bill and Jan Smet; my sister, Kate Reynolds; my in-laws, Jim and Lois Simons; and many more relatives and friends for their love, encouragement and support. I wish I could list them all.

Some of Lee's relatives: Aunt Gert, for reading and listening to these stories, and offering to tell us when Lee's recollection was "mistaken" (tar vs. wax); and Lee's cousin, Steve, for trying to explain to me what a threshing machine looked like, sounded like, and how it worked.

Loretta Hooks and Patti Jasmer, for taking the time to call and tell me more about the Kelty family, and for allowing me to use their ancestors' real names. Thank you, Patti, for the pictures of your family.

Last of all, but not least, to Harris Bennett, Lee's late husband, who passed away over a year ago, before this book could be published. I'm sorry he didn't get to see the finished product, even though he complained of hearing the stories a hundred times.

Using one of Harris's favorite sayings, I will end with: "It's as good as it's gonna get."

Right, Harris?

Foreword

I didn't set out at first to write a book. The book came to me. It all started when a friend and neighbor of mine, Leona "Lee" Bennett, asked me if I'd type some stories she had written down in a notebook.

I said, "Sure!"

I was pretty good at typing, had a computer with spell check, and I knew enough about writing to be able to edit and proofread her stories. So when Lee handed over her notebook filled with childhood memories, I was ready to go. What I wasn't prepared for was the impact her stories would have on me personally.

Being a farm girl myself, I could relate to some of Lee's experiences. She had grown up on a small, eighty-acre farm near Boscobel, Wisconsin. Her family and their neighbors were poor in money and material things, but rich in character, good morals, work ethic, and family devotion. As she grew older, Lee's memories of their lives, throughout the hard times of the Great Depression, played out one scene at a time.

Later in life, she realized what a treasure she held in her own mind. It was these treasures Lee wanted to share with her only nephew, Scott, in memory of her sister and his mother, Lenice, who had been affectionately known as "Lenny" to her family and friends. Over the course of several months, Lee diligently set about writing down everything she could remember from childhood: stories, facts, people, sights, sounds and smells. When the pages in her notebook were filled to the very last one, she handed it to me.

As I was typing Lee's stories, I felt like I was entering a different world—a dream world, in a way—even though I knew the realities of that life were quite harsh at times. I became enthralled with the characters in her stories: Lee's family and relatives; her close neighbors, the Keltys. I became a part of their world every time I sat down to type the stories. By the time I had finished, I was convinced this was material for a good book. Lee had the stories bound and gave her "book" to her nephew for Christmas. She also gave some copies to close friends.

Using Lee's real-life stories (which are written to the best of her recollection) as my "loom," I attempted to weave in characters, conversation, atmosphere and setting, and yet stay mostly on the side of legitimate biography. Some stories are more fictionalized than others. A few incidents were "telescoped" into single events, though they actually took place at separate times, to make the story more readable. Much of the conversation is my own fabrication, based on what Lee told me, because she couldn't remember exact words. However, I've made every effort to stay true to the actual spirit of the story, and to indulge in embroidery only where it wouldn't change the truth of pivotal events and exchanges.

For instance, exactly what took place when the schoolchildren rebelled against Miss Rubin I do not know, and Lee doesn't remember. Much of the conversation during cousin Margie's visit was also conjecture on my part. I did this to make the stories more interesting and enjoyable, and to draw you in so that you'd feel like you're really there. However, the basic facts are all as they appear on these pages.

Some chapters are more informational, while others are just entertaining. They are not in chronological order, except for the last two. Basically, these stories all take place while Lee was between eight and thirteen years old, and she narrates them much as she actually did when she wrote them down.

Now I'd like to introduce you to the characters in the stories, so you will understand the relationships before delving in. I hope you enjoy their antics as much as I have.

Kelly Simons
Land 'O Lakes, Wisconsin

The Characters

Leona Ahnen - This is Lee (a nickname she acquired in later years), your narrator. A tomboy at heart, a horse lover, and the self-appointed boss of little sister Lenny, she was also a trial for her mother, and her daddy's constant companion.

Lenice Ahnen – This is "Lenny," Leona's younger sister. She was the youngest Club member, and the inescapable "pest" of Leona's childhood years, an almost constant companion.

Ralph Ahnen - Leona's father, a poor but easygoing farmer. Daddy was full of love and life, generous to a fault, and was the undisputed hero of Leona's world.

Margaret Ahnen - Leona's mother, usually called "Margie" by her peers, and "Mother" by Leona and Lenny. Mother was not particularly hardy, but managed to get a lot of work done, despite often being sick.

Grandpa and Grandma Wetter - Margie's German-speaking parents; they lived about ten miles from the Ahnens in the township of Marion, and were the only folks around with electricity at the time period covered by this book.

Grandpa and Grandma Ahnen - Ralph's father and mother.

Aunt Gert - Margie's younger sister.

Aunt Margaret and Uncle Joe – Ralph's oldest sister and brother-in-law, parents to Leona's cousin Margie and her younger sister, Mary Jane.

Norman Ahnen - Leona's gopher-hunting cousin and frequent "partner in crime."

Anton Ahnen - Ralph's older brother and Norman's father.

Ernest Ahnen - Ralph's second brother, father of Bernice and Beatrice.

Margie, Bernice, Beatrice, Kathleen and Gen - Leona's cousins on the Ahnen side. Kathleen and Gen are Norman's sisters.

Arnold and Lyle - Leona's cousins on the Wetter side.

Gus and Virgil – Distantly related to Leona, and dubbed the "Toilet Paper Bandits."

Old George and Janie Kelty - The Ahnens' closest neighbors, with a language all their own.

Georgie Kelty - Old George and Janie's son, and Mamie's man of courage.

Gladys Woodard - Old George and Janie's eldest daughter, and Masel's mother.

Nellie Kelty - Old George and Janie's second daughter, tragically tongue-tied for life.

Masel Woodard - Gladys' daughter, and Leona's and Lenny's good-natured, fast-talking, best friend.

Miss Olsen – The beloved teacher at the Yellow School.

Miss Rubin – A pretty, young replacement teacher and the unfortunate target of mutiny.

The Yellow School – The one-room country school where a single teacher taught grades one through eight. Leona, her sister, their cousins and friends—in fact, anyone living within walking distance—attended elementary school there. Its name derived from its paint job, and it sat alongside a dirt road atop a big hill, from which you can see much of the surrounding countryside.

Babe and Cute – Leona's father's workhorses, a matched pair of Bays that were his pride and joy.

Topsy – Babe's daughter and Leona's first horse.

The Kelty Women consisted of, from left to right: Masel Woodard, who was Leona Ahnen's best friend and schoolmate throughout their childhoods; Janie Kelty, Masel's grandmother; Nellie Kelty, Janie's "tongue-tied" daughter who lived with her parents for her entire life; and Gladys Woodard, Janie's other daughter, who returned home after a failed marriage. The Kelty women were constant presences on the Ahnen farm, whether to draw water from their well, help out with food preparations at threshing time, or any of many other visits. This photo was taken before Masel began her annual ritual of having a permanent wave put into her hair just before school began, probably around 1935.

I could see them coming from a ways off, their straw hats moving through the wild mustard field, which was tall and ablaze with yellow blossoms: The Kelty women.

"I s'pose we best get our water for the day," Janie said, hoisting herself up from the chair.

Gladys, Masel, Nellie, Mother and I followed Janie out the kitchen door and the few strides to the well. The sky was starting to cloud over some, as I pumped the black handle up and down several times before water came.

"Looks like we might get some rain this afternoon," Mother said, looking up at the sky. "Sure could use it. The garden's awful dry."

"Don't count on it," Janie said.

The Kelty Women

I could see them coming from a ways off, their straw hats moving through the wild mustard field, which was tall and ablaze with yellow blossoms. The Kelty women were making their way to our farm for their daily water run. The Keltys lived beyond our hog pasture, down the hill, through a gate and up to the right on the far side of the next hill. They had a small farm with a garden, some chickens and, at one time, a cow. They didn't have a well of their own, so Mother offered to let them pump as much drinking water from our well as they needed.

They did have a water hole behind their house, about twelve feet across and five feet deep, which filled with rainwater in the spring. During the summertime they'd use this water for washing, along with tubs of collected rainwater. Later on, though, the water hole would become filled with thousands of pollywogs, which was loads of fun for Masel, Lenny and me, but not so good for doing the wash. Still, despite not having well water at their place, the Kelty women were immaculately clean.

Janie's ample figure was the first to appear out of the mustard field. The matriarch of the family, her housedress swayed, ballooning out when the wind would catch it just right. Janie's beautiful, white hair was cut straight across, just below her ears, and she had bangs. All the Kelty women wore their hair like this—it was a simple cut and easy to take care of—except Masel. Masel got a permanent wave every year just before school started, and would let it grow out until the next school year.

Gladys, Janie's daughter, came out next. She had dark hair, was not quite as stocky as her mother and a little taller, and wore a housedress, as well. Then Nellie, Janie's second daughter, came poking along; head up, noticing everything around her, particularly the singing birds. And finally Masel, Gladys' daughter, tagged along behind in a shirt and skirt.

Masel was my age. In fact, we were born just four days apart. We were about the same size, although Masel was a little skinnier. Her hair was brown like mine, and a bit wild-looking because of her permanent. Masel always had an infectious grin on her face.

Standing at the top of the hill, I waved my hand over my head, and Masel waved back from the bottom. All four of the Keltys were barefoot and carrying a bucket in each hand. Masel came running up the hill, easily passing the others.

"Hey, Leona!"

"Hi, Masel. You got pollywogs in your water hole yet?" I asked.

"Yeah, there's some. Cain't you come over and we'll catch some today? We can put 'em in one o' Ma's canning jars and watch 'em swimmin' around," Masel said quickly, with that ever-present twinkle in her eyes.

"Sure. Maybe I'll follow you back after you get your water, but I can't stay long. Gotta get back before Daddy comes home."

After a few more words with Masel, I saw Nellie and Gladys reach the top of the hill, then Janie, breathing rather heavily.

"Lord have mercy, this 'ralgee's killin' me," Janie said, in between gasps for air. "That hill gets longer and longer. I swear, one o' these days I'll just up and die of the 'ralgee, mark my words."

Just exactly what "'ralgee" was, I'm not sure, but I believe it was her way of saying "neuralgia," meaning to her, some form of painful arthritis.

"Your ma in the house, Leona?" Gladys asked.

"Yeah, she's in the kitchen."

The Kelty women headed for the house. Masel and I tagged along behind. Walking across the "dooryard grass," that bit of lawn just outside the door, Janie reached for the screen door first, opened it and poked her head in the kitchen. The smell of fresh-baked bread greeted our noses before we even got through the door.

"Hi, Margie. Hot today, ain't it?" Janie asked my mother, as she entered the kitchen. This was Janie's signature greeting. It could be a rainy, cool day and she'd still say, "Hot today, ain't it?" The Keltys' daily visit often interrupted Mother's work, but she never complained, even though the Kelty women were more or less a permanent fixture at our breakfast table.

"Hi, Janie. Come on in. Come in, Gladys, Nellie," Mother said politely, "And have a seat while I finish ironing these clothes."

"Don't mind us, Margie. You keep goin' with your work," Janie said, as she pulled out a chair and lowered herself into a comfortable sitting position. Gladys, too, found a chair and sat down, taking her cue from her mother and directing Nellie to the chair next to herself.

Masel and I popped in the door.

"Hey, Mrs. Ahnen," Masel said cheerfully.

The smell of baking bread, mixed with that of newly ironed clothes, was warm and inviting.

Clothes ironing was quite a project, and one of the very few domestic duties that I actually enjoyed. Most of our clothing was made of cotton, which wrinkled badly when washed, and nearly everything required ironing. First, the clothes were sprinkled with water and set aside for a while, so they were evenly dampened. The irons, usually about three of them, were put on the wood stove to heat. The handle was separate, and when an iron cooled down, you just set it back on the stove, unlatched the handle, and picked up another hot one. I thought it was great fun to iron Daddy's handkerchiefs, which I had already done for Mother before going outside to meet the Kelty ladies.

Mother answered Masel's greeting as she moved the hot iron back and forth on one of Daddy's shirts, offering her guests a piece of coffee cake left over from breakfast.

"Don't mind if I do," Janie said, helping herself to a generous-sized piece.

Janie Kelty stands with her surviving brother, John Pettit. Her other brother had been killed in World War I, and would forever remain one of her favorite topics of conversation.

Masel stepped around Janie's chair to stand directly in front of her grandmother. Wagging a finger in Janie's face, Masel said haughtily, "T'ain't whatcha want that makes you fat, it's whatcha git!"

Janie swatted Masel on the behind, and with her mouth full of cake said, "Hush up now." A few crumbs fell to the table.

"Masel, sit down, or I'll shake ye up to dogwater!" Gladys scolded her daughter. When Masel didn't sit down right away, Gladys added, "I ain't talkin' just to hear my head rattle."

I once thought that if I was given a dime for every time Masel heard those words, why, I could have bought my very own horse... and a saddle, to boot.

"So, what's Old George up to today?" Mother asked Janie, still trying to figure out what it was Masel had said. It was difficult sometimes to understand what the Keltys were saying because they talked very quickly, and with their own unique accent. The sayings they'd come up with were memorable, to say the least. Sometimes, Mother and I wondered where those sayings came from, but we didn't ask, and never found out.

"Old George's just cuttin' wood, like he always does," Janie said. "That man, he cuts and splits, and cuts and splits 'til we got us a mile-high stack o' firewood."

Cutting wood was Old George's main pastime. That, and growing crops of beans.

Daddy would plow their field for them in the springtime, using a walking plow and a two-horse team, turning over one furrow at a time. Then he harrowed the field, and from then on it was up to the Keltys to plant and cultivate their crops.

Old George and Janie would be out in that field hoeing weeds all summer in the hot, blazing sun with their big straw hats on, Janie in her usual housedress, and Old George in his bib overalls. They'd be chopping away the grass and weeds, all down the long rows of beans. In the fall, they'd harvest their dry beans. Beans will grow even in pretty poor soil, so they always managed to have enough.

"'Course, Ma, we got two stoves to feed. And the one in the kitchen ain't no light eater, you know!" Gladys reminded her mother.

Janie wiped the back of her hand across her mouth as she swallowed her last bite of cake, then shifted gears in the conversation.

"Margie," she said, addressing my mother, "Some folks can be sorely mean! Just the other day, Old George's walkin' to town, tryin' to hitch a ride..." Janie started to say.

Mother looked up from her ironing, eyebrows raised.

"Town?" she asked, thinking of the six miles he'd have to walk, one way.

"Yeah. He was walkin', and some young buck farmer came drivin' up behind him in a truck, and when the guy was only two feet or so behind my George, he lays on his horn and scares the bejibbers outta poor Old George. He about jumped right outta his pants! 'Course, you know that George's 'most deaf, so he didn't hear the car comin' 'til that son of a gun blowed his horn practically in George's ear, and about give him a heart attack!"

I had to admit that some farm humor was a little mean, but the thought of Old George jumping out of his pants was more than I could keep a handle on, and I let out a short snort. It was accidental, of course, and I slapped a hand to my mouth.

Mother gave me the 'You want a whipping? Just keep it up!' stare, with the squinted eyes, narrowed brow, and piercing look. Without saying a word, she could put me in my place.

"Who was it? Do you know?" Mother asked Janie, trying to concentrate on what was being said, instead of my rude behavior.

"You remember who 'twas, Gladys?" Janie asked. Before Gladys could say a word, Janie answered her own question. "I cain't remember who he said 'twas, but George's gettin' a bit touchy lately. Seems he cain't walk to town without someone sneakin' up on him. Them young bucks must think it's purty funny to be scarin' the old man like that. But they won't be laughin' if one of 'em ends up puttin' Old George in a early grave!"

I don't know if it was just my imagination, but it seemed the Keltys had an unusual fascination with death, and with superstitions. Knowing death was an inevitable part of life, they kind of resigned themselves to early graves whenever they needed to get a point across.

"Well, I don't blame him for getting a little mad. I think I would, too," Mother said, trying to be sympathetic.

Leaving her bread for a moment, Mother walked over to the cupboard and reached for her money jar, taking out a quarter. She walked back to the table.

"Here, Janie, is the money I owe you for this week's washing," Mother said, putting the coin in Janie's hand.

"Thanks, Margie. Those duds of your husband's, they were pretty grungy that day," Janie said, shaking her head, "But I don't mind the work. Lord knows we need the money, and washin's one thing I can do to help out."

Washing clothes was a big job, and it took someone with quite a bit of strength to do it week after week. Mother was too weak for this, so she hired Janie. Janie washed all the clothes and wrung them out by hand. Then she hung them on a clothesline, except for the overalls, which she draped over the fence in front of the house.

"Well, Ralph should have the money soon to pay young Georgie for plowing our cornfield this spring. So, hopefully, that'll help some as well," Mother said.

"That's good. We're gettin' a bit short on staples, flour and sugar," Gladys said, wanting to get her two cents in.

"That reminds me, Margie," Janie said. "The other day, the county nurse came out to our place, wantin' to see Masel. Nurse said Masel's too scrawny and needs to be drinkin' milk. So we're wonderin' if we could buy a quart o' milk from you folks. You know, our cow died last week, just up and died on us! We'd want to be buyin' some most every day if that's all right."

"Oh, I suppose. Eight cents a quart?" Mother offered, then paused a moment. "And the cream would do her good, put some fat on her bones."

Masel gave me a funny grin, lifted up her shirt a bit and pinched the skin stretched over her bony ribs. She had skin like a hound dog. It could stretch out from her body in a grotesque way, probably because she had no fat on her bones. It looked like she pulled it out a good three inches. She looked down at her bare stomach, then out from under her straw hat and grinned at everyone, showing her large front teeth.

Suddenly, Nellie gave a howl. For some reason, Masel's stretching skin struck her funny bone. And when Nellie laughed, we all laughed. It was contagious.

Looking at Nellie and laughing with her, I couldn't help but feel a wave of compassion. Nellie was born "tongue-tied." Whenever she tried to speak, people couldn't understand her, so she never went to school. Folks thought she was a bit retarded, but we knew better. The only thing wrong with Nellie was that she couldn't speak clearly.

She did have screaming fits once in a while, though. We could hear her screaming all the way over to our place. It was understandable, though. I could imagine she'd get awful frustrated, not being able to communicate. While some folks steered clear of Nellie, we knew better and just plumb felt sorry for her. It didn't help that she had kind of a strange appearance.

Ever since we had known Nellie, she'd had a large bald spot on top of her head. The rest of her hair grew as normal, and this Janie kept cut straight, just above her ears. I once saw a picture of her as a little girl, and she was cute with long, golden-blonde curls. But now those golden curls were gone, the blond had faded into a nondescript color, and the bald spot caught your attention.

My heart ached for what this poor soul endured all her life, and for the person who was denied the ability to express herself and to be understood. At least she had a sense of humor. And leave it to Masel to make Nellie laugh out loud, a rare occurrence indeed.

Mother finished the last shirt and pulled the irons from the stove, setting them on the floor to cool. Then she walked with us out to the well, casting a look at the sky.

"Well, it ain't gonna rain if you see a patch of blue big enough to make a lady a pair of bloomers, you know," she observed, repeating one of the many wive's tales that governed life in the country.

We filled each bucket almost to the top, and said our good-byes, as the Kelty women headed back home through the hog pasture and the gate.

Suddenly, I remembered about going back with Masel to her house. I started to run toward the hog pasture, quickly yelling

This photo, probably taken in 1937, shows Leona's best friend and nearest neighbor, Masel, and her mother, Gladys, during one of their daily trips to the Ahnen farm for drinking water. Masel is about twelve years old here, and Gladys would have been approximately thirty. Gladys' marriage to Masel's father had not worked out, so she moved back home to live with her parents, and brought her daughter along. Though their last name was Woodard, Gladys and Masel were nevertheless considered by the Ahnens to be inseparable members of the neighbor family group they called "The Kelty Women."

back, "I'm going to the Kelty's! Be back later, Mother!" I glanced back to see if she had heard me. She had, shaking her head and throwing her hands up in the air.

I know I was a trial for Mother much of the time. She had a hard time handling me, as we were opposites in many ways. Luckily, Daddy was out in the fields or in the barn or gone to town, and often didn't see my independent, rebellious spirit toward Mother, who was a very dependent woman. If he had known, I think he would have tanned my hide and sold it with the skins he trapped in the wintertime!

I caught up quickly with the Kelty women.

"So, you and Masel's goin' to catch pollywogs, are ye?" Janie asked. I nodded my head in agreement, unable say a word, for I was out of breath from running.

"Well, it sure's nice of your Daddy to cut a path for us to walk through this field," Gladys said, "The blossoms don't fall in our water buckets so much."

Walking through the mustard field with the Kelty women, I felt a sense of freedom, wild at heart as I was, and rebellious, always wanting to do my own thing. I stopped short to take my shoes and socks off and fling them over by a tree trunk. I intended to pick them up and put them back on just before coming home... that is, if my little sister, Lenny, didn't find them first and tell on me. I figured she was too busy playing dress-up in the house anyway. All I needed now was a straw hat.

"Okay, Club Members, it's time to come to order and have our first meeting of the summer of 1937," I said seriously, plopping down on the soft, pungent grass. Masel and Lenny sat opposite me. "Does everyone have their dues today?"

We were planning on buying a can of pork and beans some day when we had enough saved up. We were going to heat them up over a fire and eat them right out of the can, the way the hoboes all did.

The Club

Tuesday morning, a few days after school was let out for the summer, I hurried through breakfast and chores, urging Lenny to hurry along, too. Masel would be waiting for us in the hollow. I envied Masel in some ways. She didn't have any chores to do, for one thing. When I told this to Mother, she didn't pay me any mind.

"Can I take some bread and butter?" I asked Mother.

"I suppose. Done with your chores, Leona? You can't fly the coop until your chores are done."

"I finished them before breakfast," I said, feeling the sting of injustice as I cut several thick slices from yesterday's loaf. One, two slices. I laid them on the table and smeared on the butter, then did two more. Mmmmm. Nothing like homemade bread with butter. I put the slices, buttered sides together, on an open towel, brought the four corners to a peak and tied them together with a knot so the bread wouldn't fall out.

"Lenny, let's go! Are you ready?"

Lenny came bounding down the steps and caught up with me, just as I was going out the door.

Mother let us go without saying a word of warning like, 'Be back before supper,' or 'Take a jacket so you don't get cold,' or even 'Where are you going?' No, nothing like that. We just left the house with total freedom, as was often the case. No one seemed to be paying any attention to what we were doing. They really didn't have the time or energy to be keeping tabs on us. There's nothing like being a farm kid on a cool summer morning, free as the birds flying above, with the whole world to explore.

Walking outside the house, heading for the hollow, we noticed how the fragrant smell of locust trees filled the air. At the time of year when the locust trees would blossom with their delicate white flowers, our house and yard smelled like a perfume factory, but the alluring fragrance carried with it a painful secret: The locust trees had thorns that must have had some kind of poison in them, because if we stepped on one with our bare feet, the pain was excruciating.

The trees grew very tall and we always had a couple rope swings hanging around. Daddy would take the old hay ropes from the barn when they had to be replaced due to wear, and tie them high in the locust tree branches for a swing. I became very adept at rope climbing. Lenny and I would give each other a few pushes on the rope swings before heading off through the fields to the hollow.

The "hollow" was, quite simply, the valley pasture on our farm, about three hundred yards beyond Hickory Hill.

Today we were on business of sorts. It was the day our club officially re-grouped. After eight months of school, we now had the whole summer to look forward to. This club ran only during the summer months, and the dues were two pennies a week.

When we were playing this far from the well, we had to carry drinking water with us, unless it had rained the night before and there were puddles in the ditch that we could dip into. We had hidden dishes and pans behind and amongst some rocks, left there strictly for club purposes.

Tuesday morning was our meeting day, and we usually didn't miss a single meeting all summer, rain or shine. We'd move our camp around some, especially if some farmer had just disposed of a dead cow or horse in the ditch, which was the custom. The ditch was pretty close to our camp, and if this happened, we moved temporarily, but stayed in the general vicinity of the hollow.

Masel was already there when we arrived. Fifty yards from camp, I raised my hand in salutation.

"Hi, Masel!"

Lenny raised her hand, too.

"Hi, Masel!"

Chapter 2 – The Club

Lenny was five years younger than I, so naturally the responsibility of being the boss rested on my shoulders. I had to be careful what I said at times, since Lenny liked to imitate me.

"Hey, guys!"

Masel grinned back, waving her skinny arm. Masel's brown hair was cut just below her ears. Her front teeth kind of stuck out some, but she had a pretty smile. She had a clean, hand-washed dress on.

Lenny and I lived in boys' overalls. I would have worn them to church, too, if it had been acceptable, which it wasn't. Summers were definitely not for wearing shoes, so we were all barefoot. We'd go barefoot practically everywhere, except to church. There again, I would have gone barefoot, but that wasn't acceptable, either. The bottoms of our feet would become thick with calluses and stained brown from running in the dirt and manure. We didn't care. Besides, we needed those calluses to protect our feet!

"Hey, Masel, guess what I saw on the way over here?" I said, smiling and giving Lenny a wink.

"I don't know. What?" Masel asked, always gullible when it came to jokes.

"Why, I saw a dead horse in the ditch. And guess what? I one it!"

"Well, I two it!" Masel said back.

"I three it!"

"I four it!" And on went the game until Masel said, "I eight it!"

Lenny and I both started laughing. For some reason, Masel always fell for this joke, but she didn't seem to mind, and laughed at herself, as well. Masel was such an easy person to get along with. She was very compliant, unlike my little sister.

"Okay, Club Members, it's time to come to order and have our first meeting of the summer of 1937," I said seriously, plopping down on the soft, pungent grass. Masel and Lenny sat opposite me. "Does everyone have their dues today?"

Masel reached into her dress pocket and pulled out two pennies. Lenny unrolled her chubby fingers, revealing two more pennies that she had been clutching all the way from home. I dug into my pants pocket, bringing my two pennies to light. I passed a tin can around so Masel and Lenny could drop their pennies into the pot. Then I got up and put the can back in its hiding spot for safekeeping. We were planning on buying a can of pork and beans someday, when we had enough saved up. We were going to heat them up over a fire.

"Okay, guys, time to decide what needs to be done first," I said with authority.

"Hey, Leona, how about buildin' a log cabin, like we wanted to do last summer but never did get around to?" Masel suggested.

"Yeah, a log cabin, just like pioneers do," Lenny piped up.

"Yeah! It can be our clubhouse, with a door and a roof. I could chop down the trees..." I added.

"And we can make furniture outta sticks and stumps, things like that," said Masel.

"Sure, this will be fun," I agreed.

"And we'll build it all by ourselves," Lenny said excitedly.

"Okay, then I'll have to run home to get an ax," I decided, shifting into boss mode. "While I'm gone, you two can start cleaning up this place. It's a little messy from last summer."

Loping across the field toward home, I ran until I was out of breath, then walked the rest of the way. Now, the trick was going to be taking the ax without being seen, although Daddy probably wouldn't mind so long as I was the one using it. Still, I just couldn't take that chance. As luck would have it, Daddy was gone for some reason, and mother was busy in the garden.

I found Daddy's ax and quickly stole away, holding it in front of me in case Mother should look my direction. She didn't seem to notice and I ran all the way back to camp. Just as I was within twenty feet of camp, I got a sudden sharp pain in my side from running too much. As our superstitious ritual called for, I stopped and quickly looked for a rock. Finding one, I bent over from the waist, picked up the rock and spit under it, twice. When I stood up again, the pain was gone. It worked every time.

"I'll start by chopping down the trees we need, and you two can smooth out a place over there where we can lay down our logs," I said, pointing to a spot I thought would be best. Masel and Lenny set right to work, pulling up grass and picking up sticks and stones, making a clearing for our log cabin, while I set off to find a good tree in the nearby woods.

Margie Ahnen probably took this photo in 1938, of her two daughters and their friend Masel going off to their first day of school. Leona, far left, would have been thirteen years old, as would Masel, far right, and Lenny, in front, would have been about eight.

"This one will do," I thought, eyeing up a young poplar tree. I judged it to be six inches in diameter or so, and began to swing my ax toward the base of the tree.

Whump! Pull the ax out, swing it back over my shoulder and around again. Whump! This was harder than I thought it would be. *Whump! Whump!*

I kept at it like a bulldog, taking out little chips at a time, bound and determined to get this tree felled. I could see I was making headway, and finally when the cut was most of the way through, I was able to walk up to the tree and push it over, ever so slowly. Down it fell with a thump on the forest floor as I yelled, "Timber!" then de-limbed the tree and hauled it back to camp.

We laid our first "log" down right where we wanted it. I sat down on the log, and wiped the sweat from my nose and upper lip.

"Lenny, I need a drink of water," I said, expecting my younger sister to obey the voice of hard-working authority and scamper to get me some.

"Get it yourself."

"Lenny, I'm working hard here, can't you see? I'm the one doing the bull work, so the least you could do is bring me a drink of water!"

Lenny turned around slowly, dragging her feet to the water bucket, bringing back a tin can full of cool water.

Poor Lenny, I'm sure she must have felt misused at times and wished she was the older one. For the most part, I thought she was a pest, but tolerated her presence because I had no choice. I don't know how many times I heard the threat, "I'm gonna tell Mother!" and then she'd cry, but to no avail.

Still, she brought me the water.

"Mmmm, that tastes good," I said as I stood up, deciding to go get another tree. Maybe it was not with quite the enthusiasm I had started with, but I was nevertheless determined to see this project through.

Finding another tree, I began the whole process again. *Whump, whump, whump!* the ax echoed through the woods. *Whump! Whump!*

I was getting the hang of chopping down trees a little better this time around. Finally, our second tree was ready to fall, and over it went, thumph! Using the ax, I chopped off all the little branches, stepping back and forth across the felled tree. Then, picking up the butt end, I dragged the second log back to camp.

We placed this one at a ninety-degree angle to the first, forming the second wall. Then I went back for a third tree, then two shorter ones for the front wall and the doorway. Back I went again to the woods for more logs, if that's what you'd call them. As I tired, our "logs" were consistently coming back smaller and smaller in diameter, until finally I was sending Lenny and Masel into the woods to find old, already fallen trees, branches, anything that we could use.

We stacked the ends of the logs one on top of the other, staggering the corners. By the end of that day, our cabin was halfway built and starting to look like a clubhouse of sorts. Maybe it looked more like a fort, but at any rate it was ours, built by the blood, sweat, and sore muscles of Masel, Lenny and me.

I'd like to say that we finished our clubhouse, built up the walls and put on a roof; that we'd built furniture and all that we had bragged about making, but as time went on, the enthusiasm of that first day was not to be found. Instead, we settled for half-walls, no roof, rocks for our stove, pine branches for a soft bed, thin sticks lying on the ground to indicate different rooms, whatever we could get by with that didn't take too much effort. Even though it wasn't what we'd dreamed of, our clubhouse became "home" for us down in the hollow. We had good imaginations, too, and that helped to fill in the gaps.

We played "house," Masel pretending to be the mother, Lenny being the child, and I had to be the father. I'd build a fire in the fire ring outside our clubhouse, while Lenny and Masel were sent to scrounge up some wood. Sometimes, if it had rained the night before and the firewood was wet, it could take all afternoon to get a fire going. We gathered wild berries, whatever was in season at the time, and cooked them over the fire while roasting potatoes in the coals.

We each had our roles to play and chores to do, imitating life as we knew it. Masel usually did the dishes. Lenny tended to rebel against being bossed around all the time, but I was insistent and made her do whatever job she was assigned, despite her tears. She especially hated going to get firewood. Masel was always acquiescent and agreed to whatever I thought up. She was also a good mediator between Lenny and me.

One Tuesday morning, I decided to ride Babe, our big Bay, out to the clubhouse, so Lenny started on without me. I was anxious to leave, because I had a surprise to show the club members. But in my haste, I didn't tighten Babe's cinch enough, and halfway there, the saddle came loose and flopped down underneath the horse.

I was thrown, landing on my shoulder. It hurt like heck, but I never told anyone, especially Daddy. After all the times he'd warned me to pull the cinch tight, I just couldn't face him. So I suffered in silence. But when I arrived at camp with a can of pork and beans, the thing we'd been saving our dues for, I forgot all about my aches and pains. I proudly pulled the can out of my saddlebag to show Masel.

"All right, Leona! We get our pork and beans!"

I pulled out my pocketknife and cut away the top of the can. We built a fire, and when the coals were good and hot, we nestled the can down into the coals. We waited patiently, until the bean juices were bubbling and the aroma was testing our endurance of hunger. Lenny brought out our spoons and we dug in, all of us eating from the same can.

"Mmmm! This is the best pork and beans I've ever eaten," I said.

"Mmm, hmm... This is better than any food we get at home," Masel agreed, her mouth full.

"And we bought it ourselves," Lenny added.

"You know, I bet we could easily take care of ourselves if we were to run away some day and go to India, where..." I started to say.

"India? What's in India?" Masel retorted.

"Mother's cousins live there, and they're missionaries. They travel all over the place and meet new people all the time. I'll bet they see lots of wild animals, like lions and tigers."

"Can we go to India?" Lenny asked trustingly.

"Yeah, someday I'm going and you could come visit me if you want. I'm going to be a missionary just like Mother's cousins; travel the world, stalk tigers, live in grass huts..."

"But ain't a missionary someone who goes to these places to save people? You know, to tell them about the Bible and stuff? I hear some tribes worship animals and weird things like that."

"Well, yeah, I guess that's part of the job, but I wouldn't be doing that part. I'll be out exploring and hunting, bringing back meat for the missionaries, and let them do the savin' part. I'll be able to cook and eat whatever I want! I'd probably have pork and beans most every day, and wild game!"

We each helped ourselves to some more pork and beans as we sat cross-legged around the fire. Clouds were moving in and it was starting to sprinkle.

"Hey, let's get some pine branches and lay them over the clubhouse for a roof," I suggested, "So we don't get soaking wet here." We laid our spoons down on the rocks surrounding the fire, and set off into the woods for pine boughs. We each brought back an armload, layering the branches across the fort walls. Then we picked up our spoons and polished off the pork and beans while sitting under the pine roof, watching as the rain came down in heavy drops.

Our roof leaked rather profusely, but we didn't mind. This was just like camping.

The fresh smell of ozone from the rain, the smell of hot pork and beans wafting to our noses, the soft earth underneath us, pine branches above us, three friends all huddled together in their fort. Life just doesn't get any better for poor farm girls than this... at least, until we started feeling cold.

Leona sits atop Babe, who would later give birth to Topsy, the first horse Leona could call her own. The family's farmhouse is in the background, along with the 1928 Model A Ford her father bought secondhand, after several years without an automobile. Lee was about twelve years old in this photo, which means it was taken in 1937.

Sometime in June, we started mowing the first crop of alfalfa and clover, using a riding mower pulled by a team of two horses. After the new-mown hay had dried for a day, it was raked into rows with a hay rake and then hauled to the barn. We did not have a mechanical hay loader, so I drove the hay wagon along the rows, and Daddy loaded the hay in, using a three-tined hay fork.

I learned how to drive the horses and wagon at a very early age. Some of our fields were rather steep, and there was always the danger of tipping the hayrack over with a big, top-heavy load of hay.

The hayrack was like a wooden crate, nestled into the wagon frame and hanging out on both sides. It was designed this way so that if the hayrack did tip over, it wouldn't turn the wagon over, too. Because the hayrack was just sitting in the wagon and not attached, it would fall off, and the wagon wheels would stay on the ground… or so we hoped.

Hot Summer Days

"Hold 'er there a moment, Leona!" Daddy yelled.

"Whoa," I said to the horses, lowering my voice. The horses were only too glad to stop and take a break. I wrapped the reins around the standard and turned around to start tromping on the hay as Daddy pitched it into the wagon with his hay fork. This was part of my job, besides driving the wagon. I had to tromp down the hay by walking on it, back and forth, back and forth, lifting my knees high. It was just like you see those French ladies doing, stomping on grapes with their bare feet to make wine. Otherwise, the hay would be too fluffy and we wouldn't be able to get much in the wagon. By tromping on it and compacting it as much as possible, we could put quite a bit more in each load.

This tromping business was hot, dry work. The hay got down my neck, under my shirt, and was itchy and scratchy. It seemed like we were always hot and thirsty when doing this job. After a few minutes of tromping, I climbed back to the front of the wagon and took hold of the reins again.

"Move on," Daddy said.

"Gedap!" I said to the horses, and with a gentle jerk we were moving again, farther and farther up the ridge field.

After awhile I'd stop again, tromp down the hay, and then, "Move on!" Daddy would say. Finally, we had a whole load ready to take back to the barn. By this time we were at the top of the ridge field, and the road going down to the barn was very steep.

"Leona, hand me the chain, would ya?" Daddy asked, walking up alongside the wagon.

I reached down under my feet and pulled out the rusty chain, handing it over to Daddy. I hopped down from the wagon to go help as he locked one of the back wheels with the chain. The dragging wheel kept the heavy load from going too fast down the hill. Without it, the horses would never have been able to hold the heavy wagon back.

Reaching his arm through the spokes of the iron wheel, Daddy handed the end of the chain to me so I could pull it through. He then wrapped the rest of the chain around the axle and hooked it good and tight. Then we both climbed up on the wagon, and Daddy told the horses to move out.

Slowly, cautiously, we made our way down the steep hill. I could hear the wheel dragging in the back, sliding over the rocks in the road, the chain and axle binding and creaking under the strain.

"Easy now, guys," Daddy said to the horses. Babe flicked her ears, signaling she heard Daddy's warning. The sun was beating down mercilessly on the four of us. Daddy's face was streaked with sweat and dirt. I could taste the salt on my upper lip. My shirt was sweat-stained, and I could smell Daddy, too. But it wasn't a bad smell, just the pungent odor of a hard-working farmer toiling to make a living from the good earth, coaxing out of it the crops he needed to survive.

Daddy liked to remind me about the part in the Bible where it says something like, "By the sweat of your brow you will eat your food..." So true for the farmer!

Suddenly, we heard a loud SNAP! We couldn't feel the drag on the wagon anymore.

"Leona, climb back there and see what's happened."

I crawled over the mound of hay, and peered over the back end of the wagon. The chain was broken and just dangling along the rocky road. I could see where one of the links had given way.

"It's broke, Daddy, the chain's broke!"

"Well, hang on darlin', cause we're in for one heck of a ride!!"

The horses couldn't help but break out of their walk into a trot, then into a fast trot. Cute whinnied nervously to Daddy.

"I know girl, I know!"

I crawled back over the hay, down to where Daddy was driving, and grabbed hold of the standard. Visions of the field flew by us, faster and faster. The hill was too steep and we were on a runaway wagon, all tied to our poor, unwilling runaway horses.

"Hang on, Leona," Daddy hollered to me. "But if I say jump, you gotta jump, girl. Okay?"

"Okay, Daddy," I yelled back, trying to hold in the mountain of fear that was creeping up on me. All of a sudden, we hit a big rock and the wagon bounced up, taking me with it. I could feel my hand let loose from the standard and I was airborne for a moment, then came crashing back down onto the wagon.

"You still with me?" Daddy yelled.

"Yeah," I said, hanging on for dear life, my teeth gritted, steeling myself for what might come.

After what seemed like an eternity, we reached the bottom of the hill, and the road eased onto a more horizontal plane. Once on level ground, Daddy told the horses to stop. They came to an abrupt halt, their sides heaving in and out, their bodies soaked with sweat.

Daddy got down and walked up front to calm them. He patted Babe's neck and rubbed her nose, telling her everything was okay now. Then he walked over to Cute, giving her a good rub on her face, trying to ease her fright. I just stood there in the wagon, unable to move, my hands still glued to the standard.

"Leona, you okay?"

I opened my eyes a bit, peeking out to see Daddy's gentle face, hoping the nightmare was over. He was smiling.

"That was some ride, wasn't it?"

"Sure was. But Daddy, I don't ever want to do that again."

"Me, either, darlin'. Me, either. Guess I better get a new chain next time we're in town!"

Daddy climbed back up in the wagon and we drove the rest of the way to the barn.

When we got there, Daddy parked the wagon under a big door opened from the haymow in the upper part of the barn. He pulled down a huge fork hanging from a rope, and stuck it into the hay on the wagon. A long rope, pulled by the team of horses now on the other end of the barn, pulled the clump of hay up from the wagon. The fork locked into a track running the length of the barn under the roof peak, and when it was at just the right spot, Daddy tripped the fork with another light rope, dropping the hay right where he wanted it.

After all of this, someone then had to "mow" the hay, throw it back to the sides of the haymow. It was the hottest job of all, one that Daddy usually did.

A few weeks after mowing our first crop of hay, the Fourth of July rolled around, meaning firecrackers for my cousin Norman and I. We each earned money—maybe a dollar—to spend on firecrackers. Norman was two years older than I, and always bought bigger ones than I did. Some of them were at least three inches long and quite powerful. We had the most fun with some that were about two inches and made a pretty good bang. Then there were the real small ones. We'd leave these braided together and set them all off at once. We were quite creative with our firepower, but Norman was braver than I.

"Watch this, Leona!" Norman said, lighting a firecracker while he was still holding it.

"Careful! You'll blow your hand off, idiot!" I said. But Norman threw the firecracker just before it went off.

"Let's see you do it, Leona!" I shook my head no.

"Aw, come on, don't be a chicken."

But I wasn't about to take the risk, and lighted mine on the ground. Then Norman and I went to the garden tomato patch, looking for the huge green worms that we could sometimes find. They were about two and a half inches long, soft and fat.

"Aha! Found one!" I said, triumphantly.

"I found a couple!" Norman added.

We knelt down on the soft, garden earth, Norman reaching into his pocket for a firecracker. He laid the firecracker on the ground and draped the worm over it. For a split second, I felt sorry for the crime we were about to commit. This was like murder, even though it was just a worm. I felt remorseful as Norman lit the firecracker and we both stood back.

BANG! Pieces of worm flew in all directions! Norman laughed. I looked down to find a piece of green, slimy worm stuck to my shirt. I quickly brushed it off.

"Here, give me your worm, I'll do it," Norman said, with wicked delight. I reached into my pocket and felt the worm's soft body. My heart skipped a beat, and a lump rose in my throat. I couldn't do it. I just couldn't do it. Even if it's just a lowly worm, it's still a living creature, just like me, I thought.

"Huh," I pondered, trying to come up with an excuse. "I thought I put it in my pocket, but it's not there," I lied.

"That's okay, I got another one!"

Norman found his other worm, laid it over his firecracker, lit the wick, and we stepped back.

"No!" I shouted, and rushed forward to save the poor worm. In my haste to put out the firecracker, I stepped on the worm and smooshed him into the ground.

"What's the matter with you?" Norman asked sarcastically. I shrugged my shoulders as I removed my foot. There was the green worm, stuck to the bottom of my shoe.

"I don't know, it just seems kind of cruel to be doing this," I answered lamely.

"Well, let's find something else then," Norman said, and we walked out of the garden.

We eventually did find something to blow up: old tin cans. Norman's big bombs worked best for this, and I felt no remorse seeing the can go flying into the air like a rocket.

After the fun of blowing off some firecrackers, we decided to go work on our sand town, an outcropping of soft, white sand rocks near the road above our house. We spent hours blasting roads in the sandstone, and Mother's spoons made useful shovels.

One day, Norman, his older sister Kathleen and I were all playing together and somehow ended up at the Yellow School. This was what we called the one-room schoolhouse with a faded yellow paint job, where we and all of our friends attended first through eighth grade. I found that the windows were unlocked, so we boosted each other up and into the room.

It never occurred to us to do any damage, but it sure was exciting to watch a car go by on the road and think they didn't know we were inside! We did get in the teacher's desk and took some iodine tablets. These tablets were passed out to the students every Friday when school was in session, one to each student. In that time before iodized table salt, they were supposed to prevent goiters. We liked the way they tasted, almost like candy. We took only one each, but felt dreadful guilty about it ever after.

During the summer, Norman and I were death to sparrows. All the sheds and barns had sparrow nests, and the only ones spared were the ones we couldn't reach. Everyone hated sparrows. They got in the animals' feed and messed all over in the buildings, so they were fair game in most farmers' estimation. Norman and I would tear the nests down, kill the babies and smash the eggs. But in a few days, the frantic parents could always be found building another nest.

One day, Daddy bought me a Daisy™ air rifle. From then on, I stalked the birds with my gun. I didn't kill near as many this way because the gun wasn't very powerful, but it was the thrill of the hunt, now, that kept my blood pumping.

We even tried frying sparrow eggs once, in bottle caps. At one time, Uncle Anton, Norman's father, had tried making homebrew, and Norman got hold of some of the bottle caps. So we built a little fire, laid the bottle caps close to the coals and cracked the tiny eggs into them. They fried in a matter of seconds.

"Okay, Leona, you eat yours first," Norman ordered.

"No way! This was your idea, so you eat yours first."

Norman acquiesced and gulped down the burned sparrow egg with a gulp.

"What's it taste like?" I asked anxiously.

Norman had a funny look on his face. He just shook his head.

"Doesn't taste like anything 'cause I didn't chew. Although it felt kind of, um, rubbery." He looked a little pale, so I hesitated before swallowing mine.

"Go on, don't be chicken!" he chided me.

Well, I couldn't lose face in front of Norman. So, just like jumping into a cold lake all at once, I swallowed it as fast as I could, trying hard to keep from throwing it right back up again. I kept it down, but I don't think we ever fried sparrow eggs a second time.

Besides making hay, playing with firecrackers, and exterminating sparrows, for me summers also meant baseball and fairs.

Daddy loved to play baseball. He and his best friend, Orville Hoffland (whom we called "Curly"), played baseball every Sunday afternoon in a neighbor's pasture. That is, after they had removed all the cow pies.

Lots of young men had teams, and most neighborhoods had a team. They didn't need a fancy ball diamond to play in, just a level spot in a field.

Daddy and Curly had grown up as next-door neighbors, had hunted and worked together all their lives, and baseball was their passion. Daddy and Curly both played on the Boscobel team at the fairgrounds; Daddy as the pitcher, and Curly as the catcher. While the fair was going, we would go into town every day because Daddy and Curly had to play. Even Mother was interested in baseball, mainly because Daddy played.

I had absolutely no interest in watching someone else play. It was one thing to play the game, but terribly boring to just be sitting and watching. Instead, my attention was always drawn to the merry-go-round and its colorful horses. It cost five cents for a ride, and I don't how many times I wheedled that nickel out of my parents, but I must have made my Mother dizzy.

I can still hear the music, and recall the feel of the beautiful horses as they went up and down, going around and around, the whole world rushing by as I hung on. I do believe this was the start of my lifelong fascination with horses. There were a couple of seats where riders could just sit without being on a horse, but I had to ride on a horse! After a whole day of baseball, riding the merry-go-round and eating five-cent hotdogs at the fair, I always slept soundly all the way home.

During the days of firecrackers and frying sparrow eggs, Norman and I were about eight to ten years old. With a handful of matches in one pocket, a bunch of firecrackers in the other, me with one hand on my Daisy rifle, dressed in boys' bib overalls, barefoot, and hanging out with Norman, summers were fun.

I'm sure Mother worried about us, especially concerning the firecrackers, but there really wasn't much adult supervision. In fact, there was none. We were expected to be responsible and somehow, we survived.

"Brriiiiinnnnng, brriiiiinnnnng, brring, brring!"

Mother jumped up from the kitchen table to answer the telephone. We were on a party line, and our ring was two long and two short. Mother would answer, and all the neighbors "rubber-necked," or listened in. If the conversation was boring they'd just hang up, but if it was juicy, they'd silently hang in there until the last word was said.

Our telephone hung on the wall in a corner of the kitchen, just a wood box with bells on top, a crank for ringing, the receiver on a cord, and the mouthpiece that you spoke into attached to the wood box. Mother picked up the receiver and held it to her ear.

Margie

"Hello?"

Mother smiled, turned to us and said quietly as she listened, "It's your Aunt Margaret." She spoke into the mouthpiece again.

"Oh? She does?" Pause. "Great, that would be nice." Another pause. "When's she coming?"

Mother loved to gab on the telephone, but I didn't. In fact, I hated speaking on that thing. Besides, at least in our house, children weren't allowed to use the phone anyway. It had two dry cell batteries that had to be replaced occasionally, and whenever we heard a thunderstorm coming, we disconnected the outside ground wire to save it from any damage, should lightning score a direct hit on our house.

When Mother called her younger sister, Aunt Gert, who was on a different line, she rang "Central" in Boscobel with one long ring. Central would answer, Mother would give them Aunt Gert's number, and Central then connected the lines by plugging a cord into their console. Mother and Aunt Gert talked every day. I often wondered if they'd ever run out of things to say, but as far as I know, that was never a problem.

"Okay, Margaret, we'll be watching for her around noon," Mother said into the mouthpiece, then hung the receiver back on its hook. She turned to us as we finished our breakfast.

"Well, girls, it looks like we're going to have a visitor coming all the way from Stoughton."

"Stoughton?" I asked. "Isn't that where Aunt Margaret and Uncle Joe live?"

"Yes. We were there once, but you might have been too young to remember that trip. Stoughton's about a hundred miles away. It's an awful long trip. We went with Uncle Anton and his family years ago," Mother reminisced.

Because we lived far away and had had no car for a long time, we'd had very little contact with Aunt Margaret's family. Aunt Margaret was Daddy's oldest sister. She and Uncle Joe had two daughters: Mary Jane, who was a little younger than me, and her older sister, Margie. I had no recollection of the trip we made to see them, but I did remember Margie. She had made quite an impression on me.

"Well, I guess Margie is going to be driving all the way over here by herself, in a car, to visit us," Mother explained.

"Margie? All right!" I exclaimed.

Margie was our oldest cousin on Daddy's side of the family. She was ten years older than me, and had bright red hair. Margie was a girl that just sparkled with life and enthusiasm, so outgoing and full of fun. She was born with no hand on her left arm, just a couple little fingers on a stub at her left elbow. But Margie, true to character, didn't let this bother her one bit. She could do anything with her stubby fingers, and never once complained about it. Very few women drove cars, but Margie did. And she was coming all the way from Stoughton!

"So, girls, let's get to work right away and get this place cleaned up for Margie's visit," Mother said.

Lenny and I jumped right into our chores for a change, working with glad hearts as we helped Mother clean the house and prepare a bed for my cousin to sleep on. Mother and I cleaned the kitchen and dining room downstairs, while Lenny cleaned the one bedroom upstairs, making the two beds and dusting the dressers. Mother polished the black cook stove and the Hoosier cabinet in the kitchen, and made sure the washstand and basin in the corner were respectable. She even straightened up the pantry under the stairs. In the dining room, I dusted and straightened up the cot, dining table, chairs and buffet. The wood-burning heater had been removed for the summer, so I didn't have to bother with that.

Just thinking of Margie made my day. Here was a girl I could look up to. She was older than me, drove a car, and went to high school (which was a little unusual in those days). She was pretty independent, but I figured this was on account of her living in the city. I couldn't wait for her to come.

Finally, around noon, Mother, Lenny and I finished polishing every last detail of the house. Mother wanted things to look nice because, after all, she wanted to make a good impression on her niece, who was coming from the city. She didn't want Margie to think we were backwards or uncivilized. So we scrubbed and polished, beat the rugs to death, fluffed up the pillows and bed ticks, and baked bread and pies. By noon we were all exhausted, but still anxious for Margie's arrival.

We heard Margie coming before we actually saw her. "Aaa-oooooga! Ooga, ooga," we heard, coming from the road above our farm.

"It's Margie, it must be her!" I yelled, and ran out of the house to see. I saw her coming down the drive in her shiny black car, reaching forward to squeeze the horn.

"Aaa-oooooga!"

I jumped up and down, waving a hand in the air. Margie came driving right to me, stepped on the brake and came to a stop, dust billowing in her wake.

"Leona?" she asked, looking at me intently.

"Yeah, it's me, cousin Margie!"

"Well, look at you, Leona. All grown up, or almost there, anyway. My word, you've grown since I last saw you!" Margie declared.

I brushed at my clothes and stood taller, raising up just a bit on my bare feet. Oops! I suddenly became aware that I'd forgotten to put my shoes on, though Mother had asked me specifically to wear shoes. Just this once, she had pleaded. But at least I had my good clothes on, and hopefully Margie wouldn't notice my bare feet... not that she would care, I knew.

Margie stepped out of the car, standing right in front of me. She looked directly down at my feet, a big grin spread across her face. Mother was coming out from the house, waving her hand. Margie proceeded to sit down on the running board and pull off her own shoes and socks, shoving her toes into the dirt and wiggling them like a little girl.

"There, that's better," Margie said.

I grinned from ear to ear.

"Margie! So good to see you," Mother said, as she walked up and wrapped her arms around her niece.

"Good to see you, too, Aunt Margie!"

Margie's red, curly hair was popping out in a million places from under her hat. She reached up and took her hat off, allowing the bright red curls to escape their confines.

"Well, let's go inside and have a bite to eat, ladies," Mother said. We all headed for the house at once, Margie's good arm around Mother, and her short arm across my shoulders.

"Have a seat, dear," Mother said, as she pulled out a chair for our guest. "You must be exhausted and hungry."

"Well, yes, I am."

"So, how was your drive?"

"Oh, just fine. I like getting out on the road and driving for hours on end. The freedom of being able to go wherever you want to is such a good feeling. Why, I could drive across the whole country if I wanted to!"

"Would you like a sandwich, dear?" Mother asked politely, holding the plate of sandwiches in front of Margie.

"Oh, yes, I'm starving," Margie said, reaching for a sandwich with her good hand. As the four of us sat around the table, I watched every move Margie made, noticing how she talked, how she ate her food, how she smiled. Suddenly, Daddy walked through the door.

"Margie, my girl!" Daddy said, walking up to give her a big hug. Margie stood up and wrapped her arms around Daddy, knowing full well she was his favorite niece. They hugged a big bear hug, then Daddy took a seat at the table and helped himself to a sandwich.

"So, how are things in the city? You got a beau yet?" Daddy inquired, teasingly.

"No sir, but whoever I marry someday is going to have to

be at least as handsome as you, Uncle Ralph, or I won't have anything to do with him!"

Daddy laughed heartily and patted her on the shoulder.

"That's my girl! Mark my words, one of these days some young buck will come along and catch your eye. And when that happens, you be sure to bring that young man over here so I can get a good look in his eyes."

"Why would you do that?" I asked, puzzled.

"Because, only another man can look into a young man's eyes and know for sure what his intentions are. Women can't do that, they just see the outside cover and have no idea what's on the inside."

"Oh? Well, I think I had a pretty good idea of what your intentions were when we were courting, Mr. Ahnen!" Mother retorted.

Daddy choked down his last bite of sandwich.

"Of course, dear, you're an exception," he said, giving Margie and I a sly wink as he stood up to go back outside. "Gotta get back to work!"

"Say, Margie, how would like to go into town tonight?" I asked.

"Say, Leona, what's in town?" Margie mimicked.

"Well, Saturday night is the social night for farmers in Boscobel. The stores are all open, or we could go see one of those shows at the Blaine Movie Theater..." I started to say.

"Or we could go get an ice cream cone," Lenny added.

"Yeah. Whenever we go with Mother and Daddy, it's so boring, because they always stop to talk with someone we don't know. But if we go with you, it would be fun!"

"Oh? Is that so? We don't go very often," Mother said, a little testily.

"Mother, we just want to have some fun with Margie, is all! There's usually a western at the theater on Saturday nights, and ice cream cones are only a nickel. I've got some money saved up."

"Well, you better ask your father before you go running off into town."

"I'll ask Daddy at supper!"

After dinner (that's what we call lunch, on a farm), Lenny, Margie and I headed outside to a field with millions and millions of little blue Johnny Jump-ups growing everywhere. We could see this blue field on Kelty's land, on the little hill just about 500 yards or so beyond our hog pasture. Johnny Jump-ups grow only where it is sandy and sunny, so we didn't have any on our land. Lenny wanted to pick a bouquet of these flowers for Mother to put on the table that night at supper.

Holding onto Margie's good hand, Lenny skipped along beside her as I walked on the other side, watching the way Margie walked, listening to the way she talked. She was going barefoot, just like Lenny and I.

"Boy, the sun sure is shining today," Margie said, gazing up toward the blue sky. I bent my head in the same direction, squinting madly to keep the sun out.

Soon, we were engulfed in a sea of blue-violet flowers, at least a few acres wide and deep. Lenny began to pick, while Margie lay down on her back in the field, one knee bent upward toward the sky, the other leg crossed over the first. I went to lie down beside Margie, but someone coming across the field caught my eye. I waited, standing still until I could make out who it was.

"Oh, Masel!" I waved.

Masel came trudging across the field, crossing directly over to where we were.

"Hi, Masel, I want you to meet my cousin, Margie," I said, trying to lead her over to the girl lying in the grass and flowers. Margie sat up to greet Masel.

"Just a minute Leona, I gotta tell Lenny something first," Masel said seriously, ignoring the stranger on her land. I followed Masel to where Lenny was picking flowers. I was feeling a little confused about Masel's dour look, as she was usually so sunny and cheerful. Masel put her hands on her hips just before approaching Lenny. Lenny stopped picking to greet Masel.

Then with utmost importance, Masel informed her, "Lenny, you gotta wait to pick these flowers 'til I get a chance to pick all I want first. Then you can have some."

Lenny stood up straight, looking directly at Masel with squinting eyes, a puzzled look on her face. For a moment we were all silent, a bit dumbstruck. Lenny looked around at the millions of flowers, then looked back at Masel. Suddenly a fit of

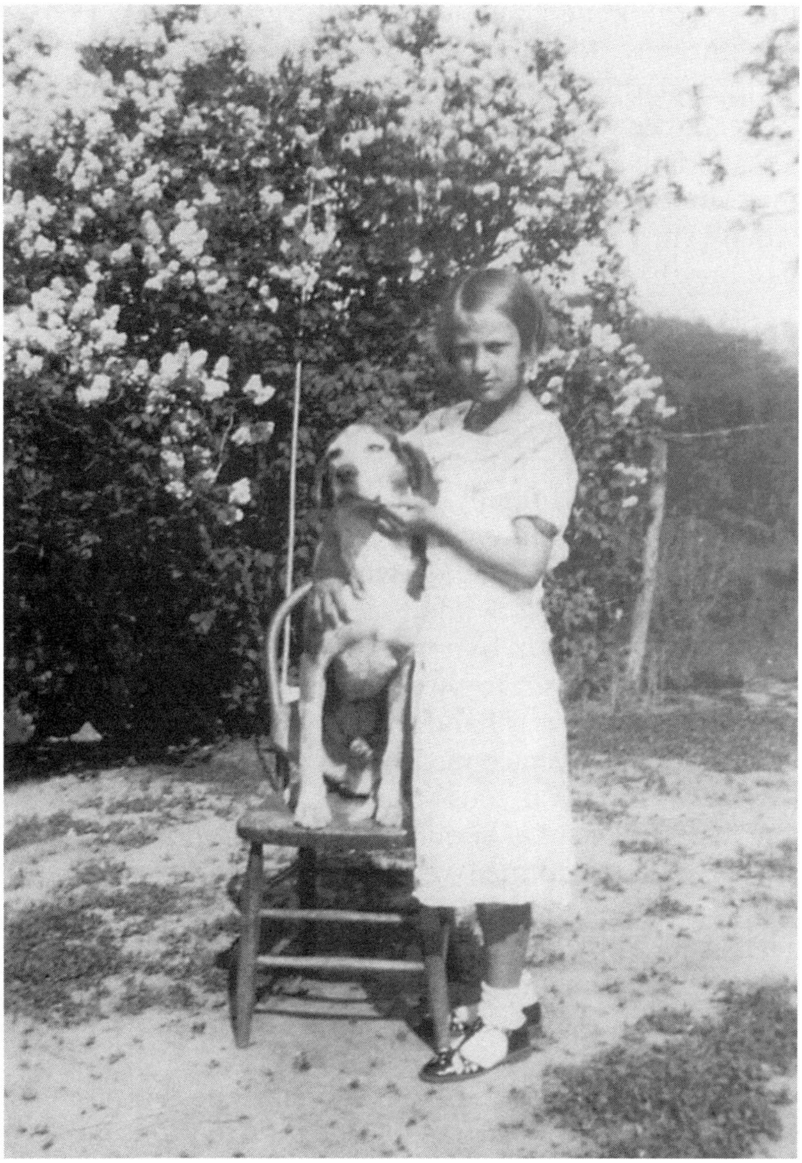

Masel, posing here with an unknown dog, was usually a cheerful, friendly girl, but Margie's arrival at first made her nervous about losing her best friends, Leona and Lenny.

laughter overtook me. I fell down on the ground, laughing and hooting until my sides ached. Masel was nonplused, hands still on her hips, staring at Lenny. Margie heard the laughing and came over to see what was happening.

"Well, Masel just informed Lenny that she can't pick the Johnny Jump-ups until Masel herself has picked all she needs first, then Lenny can take what she wants," I explained, between gasps of breath. "As if there aren't enough flowers to go around!" Then I lost it.

Margie, seeing the humor, began to laugh, too. Soon, tears were streaming down our faces, and we were falling down on the ground, rolling and laughing. All the while, Masel, still standing with her hands on her hips, kept an eagle eye on Lenny, looking a bit angry and hurt. I finally sobered up enough to go talk with Masel. Lenny was standing with her arms filled with flowers, ready to burst into tears, not knowing quite what to do.

I walked over to Masel, and Margie followed, speaking first.

"Well, Masel, we don't know each other, but my name is Margie, and I'm a cousin to these two thieves," she said, smiling. She reached out her right hand to Masel, who hesitantly took Margie's hand, barely cracking a smile. This was so unlike Masel!

"What's wrong with you, Masel?" I asked.

She just shrugged her shoulders and glanced at Margie. Suddenly it dawned on me. We were Masel's only friends, and anyone or anything that would possibly come between us might seem a threat to her. But Margie's visit was such an exciting time that I plumb forgot about Masel's feelings.

"Hey, Masel," I suggested, "Let's walk over to your place and you can show Margie your playhouse."

Masel's face slowly lost its sourness, and the sunny, cheerful girl we knew emerged at last.

"Yeah, sure. Come on," Masel said.

So, the four of us walked through the blue field on our way to the Kelty's. Margie kept a stream of conversation going between herself and Masel. Masel was finally warming up to my cousin, and found the courage to ask, "Margie, what happened to your arm?"

Margie looked thoughtfully at Masel and replied, "Oh, I was just born this way."

"Must be kinda hard doin' things. Do folks make fun of you?" Masel asked, concerned.

"You know, the way I see it, we all have a handicap of some sort. Some people have outwardly visible handicaps, like me with my arm," Margie started to say.

"I got an aunt who's got an outward handicap, too," Masel interrupted.

"Oh? What is it?" Margie asked.

"Well, she cain't talk right. It's like her tongue's tied to the bottom of her mouth. So people thinks she's a bit touched in the head, but she really ain't. She just cain't talk right, that's all."

Margie nodded her head, understanding.

"Well, like I was saying, some folks have visible handicaps that you can see, and others have handicaps in here," Margie said, pointing to her head and chest.

"What do you mean?" I asked.

"Some folks have problems with their heads, you know? They can't think straight, and they make bad choices in life, then life itself becomes one tragic event after another. And some folks have handicaps in here, in their hearts." Margie pointed to her chest. "Maybe they're filled with jealousy or anger or resentment, or just plain meanness. I'm sure you've met people like this, and they're no fun to be around. Well, that's just as much a handicap as having a bad arm! In fact, even more so."

Masel and I nodded our heads in agreement.

"With me, I carry my handicap on the outside. It's just a physical flaw, and it may not be pretty, but inside of me, I'm just as fresh and alive and beautiful as this field of blue flowers!"

Walking up to the Kelty's place, we could see a huge lilac bush beside the house. Masel's white cats were lying lazily under the bush. They were Masel's special pets, and they had to be white or she wouldn't have them.

Their yard was small but kept very neat, as was everything. The Keltys had a lawn mower leaning up against the house. It reminded me of a time they let me borrow it. I wheeled that thing all the way from their house, across the field, through our fence gate and up the path through the hog pasture, and mowed our lawn once. We didn't have a mower, so this was a new and exciting experience for me.

We saw Janie in her garden, in her usual housedress and wearing a straw hat. Janie loved to pick bouquets of lilacs for Lenny and me, and she always had a few annuals in her garden, too. There was a broken down fence around the garden, and just inside of its little wooden gate was a bed of asparagus.

"Hi, Janie. I want you to meet Margie," I said.

"Hot today, ain't it?" Janie offered her usual greeting, as she gingerly straightened up from being bent over. We filed into the garden, one by one.

"Grandma, this here's Leona's cousin, Margie," Masel said, introducing her newfound friend. Margie stepped forward and shook Janie's hand.

"I'm glad to meet you, Ma'am," Margie said, smiling at Janie.

"Oh, you just call me Janie." Turning to me, she asked, "Leona, how'd you like some of this spare grass in with a bouquet of lilacs?"

I looked down at Janie's "spare grass," which was really her bed of asparagus, already gone to seed. Janie didn't seem to realize this, or that it was edible.

"Sure, it'll look real pretty," I said.

Margie took a second look at the "spare grass," too, and she seemed to recognize it for what it really was. Janie walked over to the "spare grass" and proceeded to cut some lacy fronds to add to the bouquet of lilacs she had already picked. Janie stuck the fronds in carefully, then handed the lovely flowers to me. I gratefully accepted, sticking my nose into the small, light purple flowers and taking a deep breath. I reached over to hold the bouquet for Margie to smell.

"Mmmm. Janie, these lilacs smell better than any I've ever seen," she complimented.

"Yeah, picked 'em from over there," Janie said, pointing to a huge lilac bush standing beside the house.

"Margie, wanna see my playhouse?" Masel inquired.

"Oh, sure, Masel. Let's go! It was nice to meet you, Janie," Margie said.

"Yeah, nice to meet you, too. I gotta get off my feet for a while. My 'ralgee's killin' me," Janie said, wincing. "So, you all go ahead and see Masel's house."

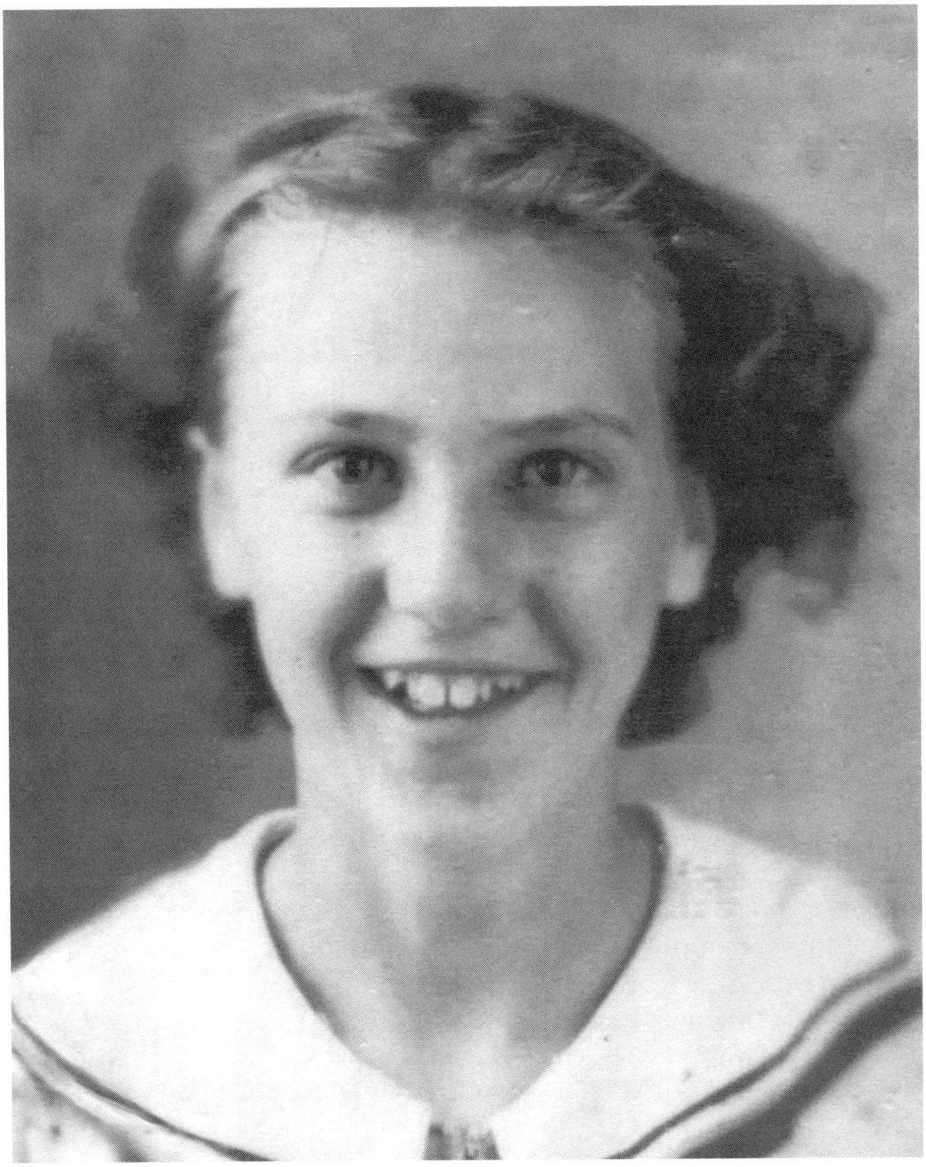

This school portrait of Masel Woodard was probably shot in 1937-38, and shows her usual cheery, somewhat mischievous expression. She would have been twelve or thirteen years old and in grade seven or eight at the Yellow School. She had clearly begun having her hair permed by this time, and was a regular member of The Club.

When we were some distance from Janie, Margie asked me, "What's 'ralgee?"

"Oh, that. It's just her way of saying she has aches and pains. I think she means "neurology" or something, which she thinks is a form of rheumatism, but Mother says it may actually be arthritis."

Margie laughed. "What a grand old lady Janie is, with her ''ralgee' and 'spare grass' and all. I love it!"

Masel didn't hear our short conversation. She had run on ahead, showing us the way to her playhouse. Masel's playhouse was nothing more than an old car someone had parked close to the driveway in the Kelty's yard, a long time ago. No one seemed to have any idea of who left it there, or where it came from, or how it got there. The thing wasn't drivable, since it had no wheels or running gear. Old George mentioned something about it being an Essex. It did have four doors, upholstered seats, and a couple windows that worked, so it became Masel's playhouse for a few years.

"Stop," Masel said, holding out her palm to us. "Let me check for snakes first. They sure seem to like this place as much as we do."

Margie wrinkled her nose and asked, "What kind of snakes? They're not poisonous, are they?"

"Mostly no, but we do get rattlesnakes once in a great while. They can be anywhere, but you hardly ever see one. Sometimes we see a big bull snake. They're just ugly. We have lots of harmless garter snakes," Masel explained.

"In fact, one time my cousin Norman killed a rattlesnake by throwing rocks at it!" I said excitedly. "I didn't actually see him kill it, but I saw the snake after it was dead!"

"Oh," Margie said, looking down on the ground and all around.

"Are you scared of snakes?" I asked Margie.

"Well, kind of. I guess so."

"Don't worry, we hardly ever see a rattler."

Margie nodded her head hesitantly.

"Come climb in!" Masel called, as she whipped a small grass snake off into the woods. Lenny and I climbed right in, but Margie was hesitant.

"You sure there's no more snakes, Masel?" she asked.

"Naw. I only found the one and I threw him way far yonder.

I can look and see if there's more," Masel offered reassuringly, as she stood up to look once again.

Margie held up her hand and said, "Well, no thanks, guys. I think I'll stay here and watch you."

I sat on the upholstered seat, looking over at Margie, realizing she wasn't infallible after all. I mean, who's afraid of snakes? City girls, I supposed.

"Could this be another handicap for you, Margie?" Masel asked seriously. Margie laughed her infectious laugh, nodding her head.

"Yes, Masel, I believe it is my one and only inward handicap—an uncontrollable fear of snakes!"

We all had a good laugh at Margie's expense, but she didn't mind. She was a city gal and knew her limitations. Maybe that's part of the reason she liked to come out to our farm, to broaden her horizons, and have a taste of life in its purest, most basic form.

Janie Kelty takes a break from picking beans or from the garden, sometime around 1935. Not a small woman, Janie was sensitive to warm temperatures, as evidenced by her inevitable greeting, "Hot today, ain't it?" regardless of the actual temperature. Despite Janie's nemeses of "'ralgee" and heat, she was a strong, hard-working woman, the matriarch of the Kelty family.

"Is that all you can remember, Leona?"

Mother obviously wanted me to go into a detailed description of the circus for Nellie's sake. As far as I knew, Nellie had never been to a circus before, but even so, I was at a loss for words. Then suddenly, I remembered something.

"Oh, I remember something funny happening, although it wasn't so funny at the time."

Green Beans and Malted Milk

We could hear old Janie coming before we could actually see her. Janie's voice carried well on such a still, summer day. Mother and I were picking string beans in the garden, having waited until after the dew had dried from the vines. Otherwise, if you disturbed the vines when they were wet, it would cause the small beans to "rust." I stood up to see Janie's straw hat weaving slowly through the tall, wild mustard plants, as she led the way down the path Daddy kept open just for the Keltys to use.

As Janie and Gladys came walking through the hog pasture, I could see they were both wearing housedresses. Some of the Keltys' clothes were homemade, just as some of ours were, but "housedresses" like Janie, Gladys, and Mother wore were not expensive to buy. Both Kelty women had aprons on, as did Mother, which they wore when cooking and working. Dresses did not suit me at all. I much preferred wearing boys' overalls and going barefoot. The only time Mother could get me to wear a dress was for church on Sundays, and even this was a struggle.

"Hot today, ain't it?" Janie said, walking into our garden.

Janie and Gladys had come to help "do beans." Earlier in the summer, we had canned peas, and now it was the beans' turn to ripen and be harvested.

"Yes, that it is," Mother answered.

At times, Janie unknowingly tried Mother's patience because she rarely had anything new to say. But Mother never said anything, or complained. Unlike her neighbor, she was not the assertive type.

Conversations with Janie usually consisted of the same three subjects: weather, health, and her brother's stint in the army during

World War I. Yet Mother took comfort in having friendly, predictable neighbors who were willing to help when help was needed.

Right away that morning, I had tried to sneak away with Daddy before Mother caught me, but she snagged me at the last minute and forced me to stay home. This was nothing new. Working with Daddy was a pleasure most of the time, except when making hay. That was a hot, itchy job. But otherwise, I enjoyed following him around and doing a man's work. For some reason, I didn't feel so "blessed" when it came to women's work, which felt more like slave labor to me.

Mother stood up slowly, straightening her stiff legs after kneeling down on the ground for so long to pick beans. She moaned a little. At one time, Mother had been a beautiful, delicate young lady. During the days when Daddy was courting her, she said, they used to write back and forth every week, even though they'd see each other on weekends.

I've seen pictures of Mother that reveal a small, oval-shaped face with large, gentle eyes, a small nose, and a thin upper lip. She wore her hair pulled into a twisted bun at the back of her neck. Her skin was smooth and pearly white, with dark, wide-set eyebrows above her eyes. Youth and beauty on were her side for a while, then time and work took their toll.

Farm life was hard. The wrinkles on her face gave away too much worry and time in the glaring sun. The calluses on her hands from years of manual labor, the aches and pains in her back from hours bent over milking, all told of the cost.

Mother was never a hearty, robust lady to begin with, but she managed to do an awful lot of work. She had a big garden and canned hundreds of quarts of vegetables and fruits using a wash boiler on a wood burning stove in the kitchen. Along with the canned goods, we stored root crops, like carrots, potatoes, and beets, in large bins. Mother also suffered from eczema on her hands, which was aggravated by milking the cows, but she always helped milk anyway.

Mother tended to be sick a lot. Sometimes she wouldn't feel well for days. Then she'd get a jar of sauerkraut from the cellar, and once she ate that, we knew she was on the mend.

"Come, ladies, let's walk by the flowers," Mother said softly, as she carried the bucket of green beans at her hip. Janie, Gladys and I followed Mother, parading past all her beautiful annuals and perennials in the yard and garden, admiring them for their delicate color and beauty: gladiolus, pansies, hollyhocks, petunias, phlox, morning glories, dahlias, and golden glow.

Mother would buy packets of flower seeds for five cents each from the L.L. Olds Seed Company catalog, and start her own flowers. Mother also had geraniums in the house. The hill beside our house was covered with sunflowers and hollyhocks that reseeded themselves year after year. After the flower parade, Mother finally led us to some chairs under the big, old elm tree in our yard, where we sat down and began "snippling" the beans. This meant to snip the ends off each bean pod and cut it into diagonal pieces.

An hour or so later, Janie took off her straw hat. Fanning herself, she said again, "Hot today, ain't it?"

"Yes, that it is, Janie," Mother agreed, resigning herself.

Mother, Janie, Gladys and I sat on the chairs with our laps full of beans, buckets full of beans by our feet, beans on the ground, beans in our hands. The smell of beans was everywhere, and we were all "snippling" away. All the while, Janie would keep a steady conversation going, even if it was only one-sided.

"Ah, Margie, it's hard gettin' old. This darn 'ralgee keeps acting up and I'll end up in my grave before I know it," Janie complained. Mother nodded her head and kept working. "I wonder if this weather's goin' to get hotter than it is today. Couldn't get much hotter, ya think?"

Mother shrugged her shoulders and said, "Who knows, who knows?"

"I ever tell ye 'bout the story my brother told, the one that was in World War I? I prob'ly have, but I cain't remember who I tell and who I don't tell. Ye heard that story, Leona?" Janie asked, her hands busy as she talked.

"Of course they've heard it, Mama. You tell those same stories to everyone you meet," Gladys chided.

Mother nodded again. "Yes, I believe we've heard it," she said, answering for me before I had the chance to say no, just to tease.

Janie Kelty was probably around forty-five years old in this photo, taken with her daughter, Nellie. About twenty years old in this shot, Nellie had been born with a speech impediment that prevented her from being able to speak clearly, though she wasn't known to have had any kind of learning impairment. Everyone referred to this as being "tongue-tied," and acknowledged the frustration that Nellie must have lived with all her life, unable to express herself.

"Oh, well then. I won't bore ye with it again, tho' it's a pretty excitin' story." And so the conversation went. Snip, snip off the ends of the beans, then slice diagonally, throw them into a bowl, grab another handful. About an hour later Nellie showed up, eager to help.

"Hello, Nellie," Mother said kindly, "Grab a chair and sit down with us." Nellie nodded her head and dragged a chair over to where we were sitting on the lawn. "Would you like to help with the beans?" Mother asked. She had a soft spot in her heart for Nellie. She felt sorry for the girl, being tongue-tied and having to live with such a frustrating handicap.

"Oh, sure she does," Janie offered quickly, "And what took ye so long, anyway?"

Nellie just nodded her head again, and Mother handed her some beans, a cutting board, a knife and an empty bowl. Nellie went right to work, paying close attention to what she was doing, snipping off just the very ends of the beans, and carefully cutting the pods into even lengths. Nellie was a good worker, some times better than Janie because she didn't talk much. Nellie did try to enter the conversation here and there with gestures and smiles. She was pleasant company. Another hour went by.

"Leona, tell Nellie about your day at the circus," Mother said, trying to switch the subject from 'ralgee and heat to something more pleasant.

"Well, I went with Kathleen and Norman—you know, my cousins? Lenny couldn't go, but I was allowed to go with them. It was pretty neat, the tigers and elephants and all the animals we saw." I stopped.

"Go on," Mother encouraged.

"There were goofy clowns, and acrobats doing all kinds of stunts." I stopped again, looking at Mother and shrugging my shoulders, not knowing quite what to tell about.

"Did you see the performers flying high on the trapeze? They were always my favorite act," Mother offered.

"Oh, yeah, we saw them too." I was silent again.

"Is that all you can remember, Leona?"

Mother obviously wanted me to go into some long-winded, detailed description of the circus for Nellie's sake. As far as I knew,

Nellie had never been to a circus, but even so, I was at a loss for words and this made Mother a little uncomfortable. Then, suddenly, I remembered something.

"Oh, I remember something funny happening, although it wasn't so funny at the time."

"Whhaaat?" Nellie tried to ask.

"Well, Mother gave me a dime to spend at the circus. There was lots of things you could buy—toys, balloons, food—but what I decided on was a malted milk. Mmmm, it was the best-tasting malted milk I ever had! So creamy, and rich..."

Now that I really had a story to tell, I was on a roll. Nellie was all ears, and even Janie and Gladys had stopped snippling to listen to my story. Mother had no idea what was coming, because I hadn't told her about this. But it was kind of funny after the fact, and I thought it would get a good laugh, so I continued.

"Then we went in to see this sword swallower. We paid five cents to go into this tent, and only the folks who paid got to see this guy." I stretched my neck up high and pretended to swallow the small paring knife I had in my hand. I gurgled and gagged a little, holding the knife just inside my mouth.

"Oh come, Leona, quit being so dramatic. And for Heaven's sake, be careful with that knife!" Mother scolded, rolling her eyes. So I stopped and continued my story.

"Well, this man plunged a long sword right down his throat. It actually went all the way into his body!"

I heard Gladys and Mother gasp.

"Really? Oh, that gives me the bejibbers just thinkin' on it," Janie said.

"That's right, it went right down into his stomach, and I could see the shape of the blade in his throat."

"All right. Go on, go on," Gladys said eagerly.

"Well, when I seen this happen, seen it go down his throat, all the way down into his stomach, it made me so sick that I started gagging," I said, grabbing my throat and making an awful noise. "Everyone was staring at me, and all of a sudden I threw up right then and there, in front of everybody! My whole malted milk came right back up and splattered all over the ground!"

"Good heavens, Leona, you didn't tell me about this before! Are you making this up?" Mother asked disgustedly.

Ignoring her, I noticed Nellie was laughing, so I turned directly to her and said, "So much for my dime and malted milk, huh?"

Nellie laughed even harder, then started to choke. Even Mother, Gladys and Janie were chuckling a little. I patted Nellie on the back rather firmly, saying, "Sorry, Nellie. Didn't mean to get you all choked up!"

This last pun—or shall I say stupid joke?—really got Mother, Gladys and Janie going. They were laughing so hard, they had tears rolling down their faces and they held their stomachs from too much strain. Once everyone had settled down, we went back to our business, and by suppertime we were done with all the beans.

Mother invited Janie, Gladys and Nellie to stay for some supper of fried chicken, mashed potatoes and fresh green beans. They eagerly accepted, and also took home a sack of green beans with them. Later that summer, we repeated our rendezvous with the Kelty women. By then, it was time to can corn.

Lenny was about five or six years old when Leona took this photo in front of their barn. She sits astride Babe, half of the team of Bays that performed the bulk of the farmwork. Note the melting snow clumps and detachable runners on the hay wagon in the background, indicating this was shot in late winter of 1935-36.

I started milking cows when I was eight years old, and Daddy started me out on one of the gentler cows who was easy to handle. It didn't take long before I was an old pro, and after milking for a few years, my forearms become quite strong and muscular. I could whip almost any girl at school in arm wrestling... and a lot of the boys, too.

Farm Chores and Horses

A typical day on the farm would start with, "Come on girls, time to rise and shine!" Daddy would yell up the stairs, "It's six o'clock!"

Lenny and I would just roll over in our soft, hay-filled tick bed, not wanting to get up. I never could figure why Daddy was always so cheerful in the mornings, so ready to greet the day. I liked mornings, but only after sunrise.

"Come on, Lenny," I'd say, giving her a sharp jab in the ribs.

She'd just moan and say, "Leave me alone."

"Fine. Hope Daddy tans your hide, then."

I'd pull on my clothes and clump downstairs. Daddy would be finishing a cup of coffee.

"Good morning, sunshine," Daddy would say to me. I'd just rub my eyes and yawn. "Well, let's get a good start on the day."

He'd stand up from his chair and head out the kitchen door, with me following in a trance-like state. On a working farm, there's never a day when you can just be lazy and stay in bed, or you'd hear it from the milk cows!

Entering the barn, I'd begin to wake up as a mixture of many familiar smells greeted my nose: hay and straw, cattle, horses, and the strong odor of manure. Daddy would walk over to the wall where the kerosene lantern was hanging. He'd strike a match and light the wick inside its chimney, then hang it back on the nail. Kerosene was another smell of the barn. We knew it was a great fire hazard, but were so conditioned to be careful that we really never had any close calls.

Oiled leather was a good smell that mixed with the odor of our workhorses, and Daddy loved those horses. When we moved to our farm, long known to some folks as "the Old Benson Place,"

Daddy owned a good team of bay horses. He also had Old Ted, the faithful horse that had carried him back and forth while he'd courted Mother, many years earlier. Old Ted was blind by this time, but I still remember the first time Daddy lifted me up on his back. I knew I'd never walk again, not when I could ride, the love of horses came upon me so strong.

One of the horses from the bay team was named Colonel, and he was Daddy's pride and joy. The day Colonel got sick and died was a sad day for all of us. Daddy just couldn't bring himself to throw the deceased Colonel in the ditch where other animal carcasses were disposed of, so he dug a grave behind the barn and buried him there, instead.

After he got over Colonel, Daddy bought a bay named Babe from a neighbor, and another the same size that he named Cute. Babe eventually had a colt we named Topsy, and this was my first horse. I started training her from the day she was born. The colt got so used to having me around and hanging on her all the time, that when Topsy was two years old, I put a saddle on her back. Once she got used to this, I just climbed on; no bucking or breaking necessary. She trusted me and I was confident of her.

Daddy also used Topsy in the field with the other horses. One day, I came home from school to find Topsy hitched with the other horses to the gang plow. She was a little over two years old and he was just working her a little, to get her used to pulling. Daddy was very gentle with the horses. He never "broke" a colt to work, he just slowly taught her to cooperate.

Come to think of it, that's kind of how he trained us, too.

Our eighty-acre farm was small, even compared to our neighbors. The barn had room enough for fourteen milk cows and four workhorses. When we moved to the farm, somehow Daddy got the money together to build a silo, and every fall we filled it with silage, fodder for the cows. Green corn was cut by hand with corn cutters, and hauled to the silo in huge loads with a team and wagon. We used Uncle Anton's silo filler and his tractor to run it. The corn was fed into the silo filler, which chopped it and blew it up to the top of the silo, where it was stored for winter. It was dangerous work, and many farmers were known to have lost a hand or an arm while feeding the machine.

In winter, it was my job to climb the ladder, which was inside a metal chute, up into the silo. I'd throw down enough silage to fill the bushel-sized basket seven times, then climb back down. Carrying the basket, I'd walk down the alley between the cows and horses, and dump the silage in the feeder troughs, one basketful for every two cows. The silage smelled a little like sauerkraut to me, but the cows loved it. Sort of a "cow salad," you could say. The horses got a small bucket of oats, and some hay.

After feeding the animals, I'd grab my milking stool and sidle up to old Bessie, one of my favorite cows. I'd start squeezing her udders from top to bottom in a rhythmic motion, causing the milk to come streaming directly into a metal bucket. If I was thirsty enough, I'd just squirt some right into my mouth.

I started milking when I was eight years old, and Daddy started me out on one of the gentler cows who was easy to handle. It didn't take long before I was an old pro at milking, which had to be done every morning and every night. You could never take a day off. After milking for a few years, my forearms become quite strong and muscular. I could whip almost any girl at school in arm wrestling… and a lot of the boys, too.

In the summertime, it was too hot to milk cows in the barn, so we'd go outside to do it. The cows were all very gentle and would stand still for this. After being milked, one of our old, fat cows liked to lie down. Lenny and I would climb on her back, and she'd just fall asleep, relieved of her load and so content.

Once all the milk was collected, it would be picked up every morning by a truck from Fennimore and taken to a cheese factory there. When the weather was hot, the milk had to be cooled. We set the ten-gallon cans in a water tank and stirred the milk with a long-handled spoon until it was cooled.

If we needed butter or cheese, the milk truck driver would bring them to us and the cost was deducted from our milk check, which came once a month. If we wanted cream, we just skimmed it from the top of the can. But this cut into our paycheck, as we were paid according to butterfat content as well as milk volume. The truck driver brought whey in the milk cans he returned to us every day, and this was fed to the hogs.

Early spring was the time when baby pigs were born. Daddy had to keep a close watch on the sows, to make sure they didn't roll over or lie on the little ones and crush them. Part of my early morning chores was to "slop the hogs," meaning to feed them. All summer, the porkers ran free in the hog pasture, and except for maybe six young sows, were all sold in the fall. A truck would be called to come out from Boscobel, and the squealing bunch of porkies was taken to the stockyard. Money from the sold pigs would be the biggest amount of cash we'd see for the entire year. They usually ran about $300, and there was never any shortage of bills to pay, which somehow always managed to use it all up.

Once a year, late in the fall when they were grown and after most of them had been sold, Daddy butchered a pig. He always had help for this job, usually one of my uncles. First, Daddy built a fire and placed a fifty-gallon metal barrel filled with water on it to heat. He then would shoot a pig in the temple with his .22 rifle. The pig would die instantly.

Most of these pigs weighed a couple hundred pounds. Two men would pull it up in the air with a block and tackle fastened to a wooden framework, then drop the pig headfirst into scalding water. This loosened its hair, which was then scraped off with a curved metal scraper. The pig would be gutted, cut up, and every part of it would get used.

Most of the meat was cut into chunks and packed in a huge stone crock, then covered with salt brine. To make sure it would properly preserve meat, we made sure the brine water contained enough salt to float an egg. The stone crock was then stored in the cellar, then whenever Mother needed meat, she sent one of us kids for it. She would soak the meat in fresh water to rinse the brine away, then slice and fry it.

All the fat taken from the butchered hog was cut into small chunks and "fried down," or cooked until the liquid lard accumulated, then packed in smaller stone jars. This lard would be used in cooking, for piecrusts, and for other purposes, mostly filled by vegetable shortening today.

Some of the meat was immediately sliced and fried, then packed in stone crocks with the lard. This was another way of preserving it. These methods created our supply of meat for the winter.

The cow in the foreground was named Jersey, after her breed, and was the only Jersey the Ahnens ever had. All their other cows were Holsteins. They once bought a pair of Holstein heifers from their neighbors, Frank and Emma Shimpach. In honor of this purchase, they named the two cows Frank and Emma.

Though normally docile animals, cows are not without temperament issues, as Leona discovered at the tender age of two years. She had walked up behind one of her family's dairy cows and, as curious young children might be expected to do, gave its tail a pull. The startled cow gave a kick with its rear leg that caught little Leona right in the collarbone, breaking it. Nothing was done to set the bone since, as when a rib gets broken, a collarbone simply has to be given time to heal.

Grandpa Wetter, my mother's father, always butchered a "beef," but we only had pork. Aunt Gert canned all their beef, and once in awhile we got a jar of this. It sure was good!

But then, there were the chickens.

We had a round chicken brooder house, and every spring we would buy a hundred mixed chicks for a dollar, from the Oates Hatchery in Boscobel. We raised white leghorns, because they grew fast and the hens were good layers.

Before we brought the chicks home, Daddy used the team of horses to pull the brooder house to a new location where the ground was clean and the grass was fresh. Then, Mother and I scrubbed it out with disinfectant, all to prevent sickness in the disease-prone chicks.

By the time they were ten weeks old, we were eating the roosters and had fried chicken nearly every day. I loved eating fried chicken, but how I detested the butchering that came first.

This was one area that was Janie Kelty's specialty—wringing chickens' necks. She would catch the bird, and grabbing it by the neck, she'd fling it around and around until the head twisted off. I could never bring myself to do this, for some reason. Instead, we had a large block of wood sitting by the barn with an ax stuck in it—a chicken's worst nightmare.

I'd have to catch the rooster first, which was not always convenient because they ran loose all over the farm. At times, I'd feel a might guilty as I lurked about the farm calling, "Here chick, chick, chick..."

However, once I caught one, I'd solemnly carry him over to the chopping block. Holding his wings and feet back, I'd lay his neck on the block, and with the other hand quickly grab the ax and chop his head off. I'd let go of the headless bird while it ran around the yard "like a chicken with its head cut off." After a few minutes, the reflex actions would cease, and I'd gather up the bird, dunk him in hot water to loosen the feathers and start picking them off. Then Mother would cut the bird up, roll the pieces in flour and fry them with lard in the "spider," a cast iron skillet.

This was a daily routine in the summer. By fall, the hens were laying. In winter, we kept them in the chicken house.

Collecting all the "cackleberries" was another chore of mine, and sometimes Lenny's. We gathered eggs every evening and packed them in an egg crate for when we went to town to trade them for groceries. When we had collected about twelve dozen, we carted them off to the grocery store in town, where the owner would "candle" the eggs. Holding each one up to a bright light, he'd look for the start of a chick in each egg. Any eggs with chicks, which indicated they'd been fertilized and weren't fit to eat, were thrown out. He usually didn't find any, but sometimes there were a few. We'd get about two dollars for the twelve dozen. Sometimes, we'd take this money in cash, but most times, we took it as a credit toward buying our groceries.

Later in the evening, after a delicious supper of fried chicken, mashed potatoes, green beans and applesauce, Daddy would say, "Leona, time to go turn the pump off."

This was the chore I dreaded most. We had a well and a pump that kept our stock tank, down at the bottom of the hill behind the barn, full of water. A gasoline engine ran the water pump. Daddy would start the engine in the early evening before supper, and then tell me to go shut it off when he figured the stock tank was full. So every evening, I'd get up from the table and walk outside, the night sky full of twinkling stars, sometimes a half moon casting its light down on the earth.

I'd walk the long distance from the barn, down the hill, and over to the stock tank. At night, the old gasoline engine would be spewing out sparks with every chug, and it scared me silly! I'd scoot under the fence, hold the valve down with a stick, and scramble back under the fence as fast as I could.

Many times, in my hurry, my back would get caught on the barbed wire, and I still have the scars to prove it. I can't remember how many times I yelled "Ouch!" trying to crawl out from under that barbed wire, swearing that I was going to make Lenny take over this job just as soon as possible.

In the depths of America's Great Depression, footwear was a luxury most farm families couldn't afford for their children. Leona shot this photo of her barefoot sister on June 5, 1936, the occasion of Lenny's sixth birthday. Lenny is standing at the end of the sidewalk, before the gate in front of the girls' farmhouse. Apparently, Janie Kelty had recently finished doing the Ahnens' laundry, as a pair of bib overalls can be seen drying on the fence behind Lenny. To help bring in badly needed cash, Janie used to come to the Ahnens once a week and launder clothes on a washboard for a quarter a day. Once the clothesline was filled, she'd hang the remaining clothes on the fence to dry.

We gasped when we saw this huge, bonafide twenty-dollar bill! It was a bit unusual to see one.

Uncle Anton handed the bill over to Daddy, and they shook hands to seal the deal. Daddy folded the bill and put it in his pocket, tipping his hat to Uncle Anton.

"Thanks brother. Margie wants to go to town this afternoon, get some groceries, maybe shoes for the girls. Can't have them runnin' barefoot once school starts in the fall."

Twenty Dollars

"All right Leona," Norman whispered as loud as he dared. "Pour the water down that hole and step back."

I lifted the heavy bucket filled with stock tank water, tipping it slowly, awkwardly over the hole in the ground, and trying to adjust my aim so as not to waste any. It was a long walk back to the stock tank, down Hickory Hill and all the way to the other end of the field.

My cousin Norman was kind of tall and lean, with dark hair and brown eyes. He had two sisters: Kathleen, who was three years older than I; and Gen, who was Lenny's age. We all played together at times, but mostly it was just me and Norman, I guess because we had so much in common—primarily a passion for gopher hunting. Norman and his family lived on Grandpa Ahnen's farm, the very place where I was born, about a mile and a half from where I lived.

"Soon's you're done pourin' that water, get outta there," Norman whispered again from twenty feet away.

Norman was a pretty good rock thrower and he had a pile of rocks sitting by his feet, one in each hand, and several bulging in his overalls pockets. He liked to hunt: gophers, rattlesnakes, birds, whatever. If it moved, it was fair game for Norman. One time, he actually did kill a big rattlesnake. I wasn't with him when it happened, but I saw the dead snake.

So there we stood, waiting for the poor, wet gophers to come popping out of the hole, Norman poised and ready to release his deadly weapon. To be honest, I'd have to say my sympathies were with the gophers. But on the other hand, I did help carry the water, I did pour it down the hole, and I would have gladly shared in the money Daddy said he'd give us for every gopher we killed. For five cents a gopher, I guess I was a willing accomplice; just one that felt sorry for her victims.

I poured the entire contents of the bucket into the gopher hole and then stumbled backwards to avoid getting pelted with stones. I ran back by Norman and we waited. And we waited. And waited.

"Well, where are those darn gophers? Why aren't they coming out?" Norman asked.

I ran up to the gopher hole, dropped to my knees beside it and peered down into the darkness. Where were they? Did they drown? No, they would have come running out of the hole before they'd let themselves drown. Had the water soaked into the soil and not even reached the gophers? I ran back to Norman.

"Maybe we should pour another bucket of water down the hole," we both said simultaneously.

"Yeah, okay. Take that other bucket and pour her down," Norman said.

I picked up the handle of the other water bucket and hauled it over to the hole and poured it carefully, my back aching with the strain of holding such a heavy thing in an awkward, bent-over position. But we were on a mission: A little sweat and pain was to be expected. The bucket empty, I ran once again back by Norman. He was standing with his feet wide apart, rock in hand, arm cocked behind his head, ready to attack. Again we waited, and we waited. Again, nothing happened.

"Geesh! Now what's going on with them gophers?" Norman said angrily.

The sun was beating down on us unmercifully. When he was hot and tired, Norman had a bit of a temper. He trudged up to the gopher hole, stomped on the ground as hard as he could with one foot, then with both feet, jumping up and down. So much for the element of surprise. He then found a long stick and stuck it down into the hole, ramming it in as far as it would go, trying to dislodge the elusive gopher family, but it was no use.

We decided the gophers must not have been home, so we tried another hole, then another and another, all with absolutely no luck. Finally, we gave up and headed for home.

On our way, we stopped at the stock tank and dunked our hot, sweaty heads into the tepid, cloudy water. There were things floating in the warm water: pieces of grass, a few dead insects, cattle drool. But this didn't bother us at all. We dunked our heads all the way into the water and came out shaking off like a couple of wet dogs. Feeling refreshed, we started the long climb back up the hill toward the farm.

Our property had some apple trees that Norman and I liked to climb. We also liked to climb up the corncrib and play inside. Why Mother never yelled at us for doing this, I don't know. Maybe she didn't know about all the things Norman and I did, or else she knew but pretended not to. Whatever the case, Norman and I were free to go, running barefoot the mile and a half back and forth between his house and mine.

One time I stepped on a nail, so Norman and I pulled it out and I ran all the way home. Mother put a poultice of bread and milk—or was it soap?—on it, and it healed up just fine. She didn't seem to worry about me. I rarely told Mother where I was going, and she never asked.

Anyway, after the unproductive gopher hunt, we climbed one of the apple trees, sat on a branch and ate an apple or two. We could see Daddy getting the team of horses ready to pull the wagon.

"Let's go see where Daddy's going," I said. Norman and I quickly jumped down and ran for the barn.

"Hi there, kids," Daddy said, as he finished harnessing Babe and Cute.

"Where you going?" I asked.

"Oh, I'm just going over to Norman's place. Uncle Anton wants to buy this load of oats."

"Can we ride along with you?" I asked.

"I s'pose, don't see why not."

Norman and I climbed into the back of the wagon and found a place to sit on top of the heavy gunnysacks filled with oats. Daddy climbed onto his seat, said "Gedap!" to the team and off we went, bouncing up the steep driveway out to the dirt road on top of the hill.

From the top, we could see many farms below us and off in the distance, like Billy Martin's place, Frank and Emma Shimpach's farm, Clarence Gunderson's, and the Bachim place. Off to the west, about ten miles, we could see the huge oak tree on the hill above Grandpa Wetter's farm.

Halfway to Norman's house, Daddy turned in his seat to ask, "So how's the gopher hunting coming along?"

"Ah, geesh. Do you have to ask? We didn't get nothin'," Norman said disgustedly. "We poured buckets and buckets of water down those stupid holes and not a one would come out."

"Well, try again another day," Daddy said, smiling. "Maybe those gophers were just gone on a picnic today."

"Well, if that's the case, then they've been on a long picnic. All summer, so far!"

Norman and I laughed at the thought of gophers packing a picnic basket, finding a shady spot by the creek, and eating fried chicken on a sunny afternoon, while a couple of nincompoops back on Hickory Hill tried to drown them out of their holes.

We never did get a single gopher all summer.

The ride over to Uncle Anton's farm was a nice break from the heat of the day, and helped raise our spirits after another unsuccessful gopher hunt. As usual, there was a wind blowing over the top of the hill, and it felt good on our hot, sweaty faces. Uncle Anton was outside by the barn when we pulled up with his load of oats.

Norman and I jumped from the wagon and ran for the little farmhouse, from which delicious, enticing aromas were coming. The smell of fresh-baked bread and apple pie filled our nostrils, making our mouths water. We walked in, letting the screen door slam shut behind us.

A second too late, Norman's mother yelled from upstairs, "Don't let the screen door bang shut!"

"Sorry!" Norman yelled back.

Aunt Blanche came downstairs.

"Would you stop yelling, please," she said, walking into the kitchen. "You kids hungry? Want a piece of bread or something?"

"How about a piece of that pie, instead," Norman suggested, glancing toward the windowsill.

"Well, I suppose, but then you'll have to forego your piece after supper tonight, understand?" Norman nodded his head and we sat down at the old, yellow gingham cloth-covered oak table, anticipating the best-tasting apple pie ever.

Aunt Blanche was a great cook. I think most farm women were, but she was at the top of my list. She slipped a knife into the center of the pie, sliding it carefully back to the edge of the tin. One, two, three cuts. She slid a pancake turner behind and under the crispy crust, coming up with a decent-sized piece of pie, apples dripping and falling out the sides. Aunt Blanche quickly slid the pie onto a plate and handed the first piece to me, the second one to Norman. We dug in with hearty hunters' appetites.

Before our stomachs had even begun to digest the delicious pie, it was gone and we were back out the door, to the barn where Daddy and Uncle Anton were just finishing unloading the oats.

"That about does it, Anton," Daddy said, as he wiped the sweat from his brow. Uncle Anton slid his hand into the front pocket of his overalls and came up with a money pouch. Much to my surprise, he pulled out a whole twenty-dollar bill.

Norman and I gasped when we saw this huge, bonafide twenty-dollar bill! It was a bit unusual to see one. One-dollar bills, of course, five-dollar bills, yes, and even a ten I'd see now and then, but rarely a twenty. Uncle Anton handed the bill to Daddy, and they shook hands to seal the deal.

Daddy folded the bill and put it in his pocket, tipped his hat to Uncle Anton and chided, "Thanks brother. Glad to see you're payin' on time. Margie wants to go to town this afternoon, get some groceries, maybe shoes for the girls. Can't have them runnin' barefoot once school starts in the fall."

"Well, okay. Guess that's it, then," Uncle Anton said, nodding.

Daddy and I climbed back into the wagon and set off for home. Daddy was in such a good mood on the way home, whistling the "Missouri Waltz" and patting me on the knee.

I looked sideways at Daddy's profile. He was a handsome man. Even though he was a poor farmer, bowed under with all the stress and strain of trying to pry a living from the hills of Boscobel,

Wisconsin, he had an easy way about him that allowed folks to feel welcome and comfortable when they were around him.

Daddy was a happy-go-lucky guy, not a worrier by any means, and very generous. He could make even the worst sour-puss smile on a rainy day! On the tall side, with dark, thick hair, he had a long, straight face with a nose of the same proportions. He had dark, gentle eyes, and his ears did stick out some. Daddy was a man at ease with himself and with all of life, particularly when he had a twenty-dollar bill in his pocket to bring home to his sweetheart.

"Can I see the twenty-dollar bill?" I asked, feeling rather giddy. Daddy reached into his pocket and pulled out the neatly folded bill, and handed it to me while keeping his eyes on the road ahead. I opened it up, felt the thing all over, examined every inch, even smelled it.

"Kind of smells new, don't you think?" I said.

"Mmm-hmm," was all Daddy said, nodding his head, smiling.

I tried to imagine what I could buy with twenty dollars. A saddle, maybe? I didn't have a saddle, only rode bareback, which made Mother nervous, especially when I'd ride at a full gallop down our steep hills. But it wasn't mine to spend, and we needed food before we needed anything else. I gave the money back to Daddy, who then put it back in his front overalls pocket with one hand, holding onto the reins with the other hand.

The ride home was uneventful, quiet and peaceful, listening to the rhythmic sounds of the horses' hooves clumping along on the hard-packed dirt road. A little bit of dust rose behind us, temporarily marking our trail.

Once we got home, Daddy unhitched the team while I went into the house to change my duds. Mother and Lenny were already dressed in their good clothes. They'd been waiting for us to come home. I ran upstairs to my room. Well, I should say, to "our" room. We all slept in the one room upstairs. Our house wasn't big enough for each of us to have our own. I pulled off my dirty gopher hunting pants and shirt, and threw them on my—er—our bed.

Lenny and I shared a bed, just a tick filled with fresh straw from threshing time, covered with one heavy blanket that

Mother made out of wool from sheep she'd had years ago. Mother had sent the wool away to be cleaned and made into fillers for our quilts. She covered the wool with flannel, which made a good, heavy quilt to keep us warm in the winter. Our old house could get pretty cold in the wintertime.

It was a cold trip when you needed to relieve yourself, too. We had no bathroom in the house, just a "two-holer" facing the hog pasture in the backyard, and Sears catalog pages for toilet paper. (Now, why there were two holes, I didn't quite figure out, except maybe some folks thought they might have company once in a while. I sure wouldn't let anyone in while I was in there.) Anyhow, during the winter, you learned how to run out, squat down, do your duty and run back to the house all in thirty seconds or less.

The house was wooden frame, with lath forming the inside walls and wooden siding on the outside. There was no insulation. No wonder it got so cold! But the quilts Mother made kept us toasty warm. Snuggled down into our soft, straw ticks, we'd sleep like babies through the night, not wanting to get up the next morning and put our warm feet on the ice-cold floor. During summer, though, we had the opposite problem. That's when we'd sleep outside, under the stars.

"Leona?" Mother yelled from downstairs. "We're waiting for you!"

I quickly put on my good overalls and shirt, and ran back downstairs.

"Ok, I'm ready," I said, slightly out of breath, excited to be going into town.

Daddy came walking into the house after tending the horses. He leapt up the stairs two at a time, quickly changed his duds, then came back downstairs, whistling merrily and stopping to give mother a kiss on the cheek. Daddy walked over to the water pitcher in the kitchen to clean his face and comb his hair.

"I'll be ready to go in a minute, girls!" he said, giving us a wink as he opened the door and left for the barn.

Not even five minutes had gone by when, all of a sudden, Daddy came storming back in the house. Without saying a word, he ran upstairs. Mother, Lenny and I could hear Daddy's rapid footsteps on the floor above us. It sounded like he was tearing the bedroom apart! We could hear clothes falling to the floor and some muffled, angry words.

Then he came back downstairs and started picking up things in the kitchen, looking underneath, and putting them back down, rather hard. He looked under the kitchen table, moved the chairs, and looked in the money jar. Then he turned to the living room, moving furniture, picking up cushions, throwing them back into the chairs. Without a word, he then headed straight out the door, letting it close with a bang.

From the kitchen window, we watched as Daddy climbed up into the wagon, looked under everything, then jumped out.

"Dadburnit!" we heard, all the way into the house. Mother and Lenny and I all looked at each other with shocked expressions. We knew something was very wrong for Daddy to cuss like that. Normally he was a man of decent language.

"What's wrong with Daddy?" Lenny asked Mother.

"I don't know. I really don't know," Mother said, as she watched Daddy out the window. "I wonder, did he..." she half whispered, talking to herself.

Mother finally walked outside and talked to Daddy. We heard Daddy's voice loud and clear above Mother's, as Lenny and I poked our heads out the door.

"Margie, I've looked everywhere! In the house, in the wagon... I'll go look in the barn."

Then he was gone. Mother followed him into the barn.

"Come on, Lenny," I said, and led the way outside. We walked quickly out to the barn, to find out what was going on. We carefully walked up to the big barn doors and peeked around inside to see Mother and Daddy searching for something. Now, Mother was in just as much of a panic as Daddy was.

We ventured a little closer to the scene, where Daddy was obviously becoming angry and frantic. He had picked up a bucket and was scanning the floor, pushing the straw out of the way with his foot. Then, not finding what he was looking for, he threw the bucket down. He kept pushing straw aside with his foot, swiping his leg from left to right, taking a step forward and doing it again. What was under the straw? All of a sudden, Daddy gave up, turned around and looked straight at me. I froze.

"Leona. Leona, darlin', did you give that twenty back to me when we were driving home? Or—please, please, PLEASE—tell me you kept it," Daddy begged, rushing towards me, kneeling down on the dusty, dirty floor in his good overalls and grabbing both my arms in a tight grip. He looked straight into my eyes.

Sudden panic flashed through my body and I felt my cheeks becoming hot and flushed.

"Daddy, did you lose the money?" I asked, timidly.

Daddy's head sank to his chest even though he still held on to my arms. He lifted his head back up slowly, looking once again into my eyes, but this time with a look of defeat. He nodded his head as he let me go.

I backed away slowly, with a stunned look on my face, and dashed for the house. I ran upstairs to the bedroom as fast as I could, picked up Daddy's overalls off the floor, reached into every single pocket, desperately searching for the money that I knew he had stuck into his pocket. I pulled the pockets inside out. It wasn't there. I started to turn the house upside down, looking for this money. Nothing.

I ran outside, looking on the ground, and by then Lenny was looking with me, and so was Mother. We scanned the ground all over the place, praying we'd catch a glimpse of the green bill.

Daddy had given up and was sitting outside the barn, slumped against the wall in the dirt with his good clothes on, head resting in his hands.

Mother and Lenny and I continued our vigilance on into the evening, as we changed back into our work clothes to do chores, as we made supper, as we cleaned up the kitchen, as we undressed for bed, as we climbed into bed and said goodnight.

But we found nothing. Nothing the day after, nor the day after that.

We all had to give up on ever finding the lost twenty-dollar bill; the bill that was to have bought groceries, maybe shoes for Lenny and me, the bill that would haunt us forever. It was one of the great tragedies of my childhood.

I still dream of finding that money.

Leona took this photo around 1936, looking down the road toward "The Old Benson Place," which became her family's farm when her father and mother bought it. Look closely at the lane in the foreground and you can see Lindy, a Blue Tick Hound dog, trotting along ahead of Leona. The pole sticking out of the road bank to the left carries telephone wires down to the farmhouse on the far left. The buildings at center, behind the trees, are the chicken house, then the car shed, and the granary. Behind those lies the machine shed. To the right of those buildings and further to the front are the pump house, and the round brooder house. The silo, and the barn sit to the right of those, and back between the machine shed and the barn are old piles of straw, or hay stacks. The dome-shaped pile on the far right is a haystack. The house way in back is where the Edgecomb family lived.

Reaching down to pick up the bottom edge of her apron, Mother wiped the sweat from her nose and forehead. It wasn't even eight o'clock in the morning, yet the heat of the summer day had already started to penetrate the house. Mother had been baking pies since six that morning in the wood cook stove.

"Leona, go get Lenny so we can get going on the cakes," Mother ordered.

I quickly ran out the door into the backyard and found Lenny. The Kelty ladies were walking up through the hog pasture, coming to help prepare the noon meal.

"Come on Lenny, we gotta help Mother. The threshing crew will be here soon," I said.

Lenny reluctantly got up from playing with her paper dolls and we waited for Janie, Gladys, Nellie and Masel to catch up.

Threshing Time

"Hot today, ain't it?" Janie greeted us. "I suppose Margie's been up since dawn, working on pies."

"Yeah, and now I'm supposed to start working on the cakes," I complained, as we all walked toward the front door.

It was a hot, steamy August morning. It was so humid, the air was actually kind of hazy. The threshing crew was on its way. As with most visitors, we could hear them coming before we could see them. But soon I spotted the steam-powered tractor and the thresher machine lumbering across Billy Martin's field.

Because the tractor and thresher had gigantic iron wheels, they couldn't come down the road. This would have damaged the unpaved dirt lanes, so they detoured through the fields, instead.

The minute I walked in the door, Mother gave me instructions.

"Leona, start measuring out six cups of flour. Just put it in the big wooden bowl for now," she said. "Oh, hello Janie, Gladys. Glad you could come. Certainly could use the help."

"Oh, no problem. You want me to start peelin' the spuds?" Gladys volunteered. Mother nodded her head and handed a big bowl of potatoes to Gladys.

Finding the flour sack in the pantry, I dipped my hand down into it with the metal measuring cup, bringing up a cupful of fluffy, white flour. I dumped it into the bowl, and dove back in for the next.

"One, two, three..." I counted irritably. I could hear Janie's voice above the others, taking her orders from Mother, then delegating the work to Masel and Lenny. Nellie usually just sat in the corner and watched, unless Mother gave her a simple job once in a while.

This was terrible! I didn't want to be in the house doing women's work. I wanted to be outside with Daddy and the threshing crew. I liked the guys who came with the machines. They were all neighbors.

The threshing machine was owned by Fred Updike, and the tractor by Ervin Walker. They made the rounds of the neighborhood, threshing the grain of each farmer in turn, being paid so much per bushel of grain threshed. Every farmer furnished at least one team and wagon, and they helped at every farm until the job was done, regardless of the size of the harvest. Ours was the smallest job to be done, only taking a day, but some farms took three or four days, and Daddy helped on these bigger farms for as long as it took.

I could have kicked myself for getting stuck baking cakes, instead of helping Daddy like I'd wanted to do.

"Three, four, five," I continued counting, dumping the flour into the bowl.

Mother was going to make a couple cakes for the threshing crew. Boy, could they eat! I wished I could eat that much, but my stomach would only hold about a quarter of what those men could eat. Daddy had brought home a huge beef roast from town the day before, but by the end of the meal, there would be nothing left of it, not even enough for the hound dogs.

"Oh, am I glad it didn't rain today. Thank you, Lord," Mother said to herself, as she stuck the roast in the oven. "It's just nerve-wracking, buying that much meat, baking all those pies and cakes ahead of time. Then waiting, hoping it won't rain and that we go on with the threshing as planned. I don't know what I'd do with all this food if we didn't thresh," she fretted.

Janie reassured her it would get put to use, that they'd take some if need be. But there was no need to fret. The meat was in the oven, the pies were already done, the sun was shining... and I was so mad I could spit.

"Six!" I exclaimed, finishing the flour.

"Good. Now get the sugar and the baking powder and the eggs. Oh, and Leona," Mother added hurriedly, "Go get some eggs from the henhouse, right away."

I glanced at Mother to be sure it was really her. She didn't normally bark out orders. She was usually a pretty mild-mannered lady, but this was no ordinary day. She'd get so uptight and nervous on threshing days. I suppose trying to feed a dozen men and all of the kitchen helpers would put a lot of stress on any person, but I still didn't think she needed to be so impatient.

Still, I took my opportunity and escaped from the kitchen for a few moments to get the eggs, as I had been ordered to do. As I got close to the henhouse, I could see the threshing machine rolling across our field. Daddy was by the woodpile, loading the back of the wagon with firewood for the steam engine. I glanced back at the house for a moment, then thought, "Naw, Mother won't miss me for just a few minutes. I'll come right back."

So I took off running across the field to meet the threshers. Waving my hand and yelling "hi" to the men, I ran right up to the tractor.

"Can I drive, Mr. Walker, just for a bit?" I shouted over the noise of the tractor.

Ervin Walker gestured with one hand to come on up, and he stopped to let me on. I climbed onto the tractor and grabbed the steering wheel with both hands, blood pumping so hard through my arteries I could hear it beating in my head. Mr. Walker slowly pushed a lever and we lurched ahead.

"I'm driving! I'm driving!" I yelled to Mr. Walker. He just smiled.

The tractor crawled at a snail's pace across our field. I glanced at Mr. Walker and he gave me the "OK" sign. I just kept my eyes on the field ahead, my head raised high, feeling like the tractor was carrying me majestically over the field toward my castle in the sky, like I was on a giant chariot.

Aaahhh, the feeling of power. Being behind the wheel of such an enormous machine, the wind in my face, the heat waves coming at me. The freedom, the power, the...

"Leona!" Daddy's voice suddenly boomed above the noise of the tractor. I was jerked back into reality all too quickly. Where did he come from? "Leona, get down! Your Mother wants you in the kitchen NOW, girl!"

My blood still pumping, power and freedom now turned to cowardice and fear. Mr. Walker stopped the tractor and I jumped down right in front of Daddy, feeling terribly embarrassed and sheepish.

He just pointed his finger toward the house and didn't say a word. He didn't have to. I was in trouble and I knew it. Mother must be furious. How long had I been gone? Couldn't have been too long, a few minutes at the most. On my way back, I stopped at the henhouse to finish what I had been sent to do.

"Leona!" Mother hollered from the kitchen door, "Don't bother with the eggs. I got them myself!"

Oh boy, I really blew it. But maybe, just maybe, it had been worth the trouble to be able to drive the tractor just once.

Opening the kitchen door, I saw Masel standing next to the table, helping Mother with the cakes. She was pouring the batter into baking pans. Guess I'd been gone longer than I thought.

"Leona, where have you been? I sent you out to get some eggs. Just a simple task, my dear. I had to get these cakes in the oven and you weren't bringing my eggs! Where did you go?" Mother demanded.

"Well, I just thought..." I started to say.

"You thought? You thought what? I don't think you were thinking at all!" Mother turned her back on me.

I looked over at Lenny. She had the sneakiest grin on her face.

"Lenny, wipe that smirk off..." I started to say, but was interrupted again.

"Leona, you leave her alone. She wasn't the one who ran off when I needed her the most. Of all the days you pick to disobey me, this just beats all. But I don't have time to argue right now. I want you to start peeling carrots, the whole bunch of 'em. Now!"

Spying the peeler and carrots already laid out on the table for me, I picked it up and began the long, laborious chore. Janie was looking at me, shaking her head in dismay. Masel was just trying to hide her grin.

I could see the threshing crew out the kitchen window. They were standing by the barn now, talking over the plans for the day. Oh, how I wanted to be a part of what they were doing! It was times like these when I wished I'd been born a boy.

All the while I was peeling carrots, I kept a close eye on what was going on out in our field. The field itself was littered with "shocks" everywhere. The shocks of grain had been made in the weeks prior to threshing.

By August, weeks ago, the oats were ripening, and it was time to get the binder from the machine shed. The binder was pulled by four horses. It cut the oats and tied them into bundles, or sheaves, which were dropped in rows in the field. Then the sheaves had to be stacked into shocks to keep them dry and off the ground until threshing time. A shock was made up of six grain sheaves leaned up against each other, with one more laid over the top as a cap to shed water when it rained.

I helped shock the grain with Daddy, usually in the evenings after chores were done. But when it came time for threshing, I was needed more in the kitchen than out in the field—one of the true tragedies of my life as a child.

The steam tractor and thresher were now parked near the barn. The thresher was connected to the tractor's flywheel by a long, wide belt. The spinning flywheel turned independently of the tractor's drive wheels, so the tractor could be stationary yet supply power to the thresher, turning fast-moving belts and pulleys.

The whole set-up was a bit fragile and complicated, and it took some skilled and knowledgeable men to operate it. Mr. Walker and Mr. Updike had to constantly monitor the machines, oil the bearings, adjust the belts and the sieves. Actually, it was dangerous work, and we had all heard stories of someone who'd lost a hand or an arm because a piece of their clothing got snagged on some of the moving parts.

From the kitchen, I watched as Mr. Walker fed more wood to the burner on the tractor, and another man hauled water for the engine. It took a while for the tractor to build up enough steam pressure to be able to run the thresher, but by nine o'clock, they were ready to go and the whistle blew. Now the work really began.

Facing into the wind, the three Ahnen workhorses eye the camera before being put in the traces for another day in the fields around 1937. At far left stands Cute, one half of the matched pair of Bays that Ralph was so proud of, at five years old. The other half of the pair, Babe, was a year older, standing at far right, and in the middle stood Topsy, Babe's daughter, at two years old.

Some of the men drove a team and wagon slowly along the rows of shocks in the field, while others on the ground walked alongside, tossing the sheaves onto the wagon with pitchforks. When they had a full load, they'd drive back to the thresher. A couple men would stand on the piled oats and throw the sheaves onto the feeder, an adjustable hopper through which the grain and straw passed into the machine. The hopper could be raised up or lowered down to adjust the flow of the sheaves into the machine.

Inside the thresher, the oats passed between a spinning, toothed cylinder and a grooved plate. The cylinder whipped the grain onto the grooved plate, knocking the kernel from the head without crushing it. From there, the grain was raked onto a coarse grille, which separated most of the straw from the grain. The straw was blown out a chute and onto the ground, making a straw stack that grew into a mountain by day's end.

Meanwhile, small chaff and grain passed through a series of progressively smaller sieves. Finally, a blast of air from a blower blew away any remaining chaff, leaving the heavy grain that would then fall into a hopper, which led to the ground. Men would hold gunnysacks underneath this hopper to catch the grain, tie them shut with heavy twine and throw them onto a wagon. When the wagon was full, it was taken to the granary in town.

All the men had jobs to do, and if done right, it was a pretty smooth-working operation, but also a dusty one. The machines and the men were covered with chaff. It got in their ears and up their noses. Some of the men even wore their handkerchiefs up over their noses to keep from breathing in so much dust. Lenny was told to take a jug of lemonade out to the thirsty men. I was green with envy.

"Oh," I almost cried to myself.

The tractor, my moment of glory, had faded all too soon, leaving me stuck here in the kitchen. Of all the fates I could think of, this was the worst, almost like torture. But I didn't have much time for a pity party before Mother had me on to another job.

Before we knew it, we heard the whistle blowing from the steam engine, telling everyone it was noon, time to stop and eat. The men parked their horses and wagons by the barn, unhitched the horses, gave them some water and hay and an hour's rest.

The kitchen smelled delicious, of baked apple pies and spice cakes, roasted meat and fried chicken, homemade bread with lots of butter, mashed potatoes, green beans, cooked carrots, sliced tomatoes right from the garden, and a large crock of fresh-squeezed lemonade. Mother sure did do it right when it came to feeding the threshing crew, and they knew it.

Sitting under our big elm tree in the front yard, the men waited, grateful for the cool shade. They were all dressed similarly, in blue work shirts and bib overalls with a red handkerchief in their back pocket or tied around their necks. They wore straw hats, and their hands, forearms and faces were burned brown by the summer sun; except their foreheads, which were white from the hat line up. I always thought they looked kind of funny when they took their hats off. Daddy, and I'm sure the other men as well, had this perpetual "farmer's tan."

They were all hot and sweaty and dusty. Mother had put several washtubs full of clean water, a couple washbasins and plenty of Ivory soap in the yard for the men to wash up with. Once finished, they'd throw the dirty water on the ground.

There wasn't room for all to eat at once, so half ate while the others sat and visited, then the other half ate at the second table. The men devoured everything in sight, right down to the last chicken wing and fresh tomato slice.

Mother always saved some of the food back in the kitchen for herself and the kitchen staff, but we couldn't sit down to eat until all the men were fed and had gone back to the field. Then we fixed plates for ourselves, and sat outside under the shade of the old elm. This was the part we had all been waiting for: eating the good food our farm produced and our own hands had prepared. Everyone was pretty quiet as we ate our meal. There were just the sounds of forks scraping plates and a few complimentary words about the food.

Mother seemed to have forgiven me. At least, she wasn't cross anymore, so I took a leap of faith and asked, "Mother, after we're done cleaning up the kitchen, can I go out in the field? Please?"

Now that the meal was over, Mother seemed more relaxed and in a generous mood. She said "Sure, go ahead, but only after the kitchen is cleaned up."

I could accept that, I guessed, and got up from our dining area under the tree to get started on the big mess in the kitchen. The rest of the ladies were in no big hurry, even Masel. She just couldn't understand what it was about threshing time that drove me to want to be out in the hot sun, working with all the men.

Kitchen chores done, I could hardly wait to get outside. Lenny followed me out to the field as the Kelty women were heading for home. We found Daddy and asked what we could do to help. Daddy rubbed his hands over his face, smearing the sweat and dust all together. He walked over to the water bucket and dumped a scoop of water over his head. Drying his face with a dirty towel, he said, "Well, girls… Lenny, you can see that the men get some water to drink. And Leona, maybe you can help me build the straw stack. Okay?"

The dirtiest job at threshing time was building the straw stack, and usually the farmer whose threshing was being done, along with a helper, did this work. The straw, left over from the threshing machine process, was blown out a chute onto the ground, and someone had to build a stack from this straw that would stand tall and shed rain and snow. It would later be used as bedding material for the animals throughout the year.

Building the stack was hot, messy work and wasn't exactly what I'd had in mind when I offered my services. But as luck would have it, old Mr. Walker saved the day.

"H-hey, Leona!"

I looked to see who had called my name.

"Oh, Mr. Walker!" I said, raising my hand in greeting.

He motioned for me to come on up, but I held up one finger to let him know to wait a minute. I ran over to Daddy and asked, "Can I go help Mr. Walker with the tractor instead? He asked if I'd come on up, Daddy."

Daddy nodded his head and I ran back to the tractor, climbed up onto the back, and found a place to stand. It was very hot standing close to the burner, especially when Mr. Walker opened it to throw some more wood in. I looked at all the levers and pointed to one, asking, "What's this for?"

"D-don't p-pull that one or-or we'll s-s-start h-heading west! D-don't th-think your P-pa w-would a-ap-preciate that."

I looked at Mr. Walker's face and grinned. I liked Mr. Walker, stutter and all. He was a man of small stature, kind of thin and lean with grayish brown hair, dark complexion, lots of wrinkles on his face, and big brown eyes that were so kind. He had on a straw hat and some old overalls covered with grease. There was a fine layer of dust all over him, and over everything, from the threshing machine.

"Turn 'er down a minute, Ervin!" one of the men yelled.

Mr. Walker moved a lever and looked forward to see what was happening. He hopped off the tractor to go see what the problem was. I jumped down, too. Maybe I could help or do something.

"Eh, the dern thing's jammed in there," Fred Updike said, disgustedly pointing to the thresher. "We gotta stop for awhile, so I can get in there an' unjam it."

We watched as Mr. Updike worked on the machine and finally got it in running order again. It was expected that things would go wrong once in awhile, but Mr. Updike was a very capable man, and he knew how to fix anything.

I ran back to the tractor, climbed up and stood off to one side before Mr. Walker climbed up after me.

"Okay, Ervin! Give 'er some power!" Mr. Updike yelled.

Mr. Walker slid the same lever back the other direction and the steam-powered engine began to chug. The thresher started to make its "whooshing" sound, and the men went back to work once again. Wagons were coming and going all day long. We had a good crop that year.

Toward early evening, things started to slow down. Finally, the threshing was done, and someone hauled the last load of grain to the granary. The men all gathered around Mr. Updike as he explained whose farm was next, the following day.

Daddy could be gone for days, if the farm was pretty large. Mother was always curious about what the other women were serving to the threshing crew on neighboring farms. Daddy was kind of critical of some places where he ate dinner, but overall, the women did a fine job of feeding the hungry men.

Lenny came around with the water bucket and offered all the men one last drink. She started to go around me, and I reached out to grab the bucket before she passed me up.

"Lenny, I need some water too," I said, picking up the dipper and taking a long draw. Doing men's work was hard stuff, and it sure gives one a powerful thirst. I drank some more.

"Leona, I gotta move on," Lenny said impatiently.

I put the dipper back in the bucket and let her go. As I swallowed my last gulp of water I ended with a long, drawn-out, "Aaaahhhhh," just like the men did after taking a drink.

Mr. Walker looked over at me and winked. I grinned back at him, feeling very important and special.

Now THIS was men's work. No baking cakes for them! No peeling potatoes or cleaning up kitchens. They did real work, or so I thought… the kind I liked to do, anyway.

The Kelty family, like many farm families, had three generations living in the same house. They lived across a hog pasture from the Ahnens. Without a telephone or automobile, they were quite dependent on the Ahnens for help in an emergency. Standing in this 1937 photo are:

At far right, Old George, who was rather deaf; his wife, Janie, to his right; and their granddaughter, Masel Woodard, in front. To Janie's immediate right is their daughter Nellie, and to her right is Masel's mother, Gladys Woodard. In back, behind Gladys and Nellie, is Gladys' husband Chester Woodard, and Chester's son stands in front with his half-sister, Masel. When Gladys and Chester were divorced later, Chester and his son moved to Iowa.

"And what about the convict? Where did he go?" Mother asked, glancing toward the fields.

"Him? He musta hightailed it into the woods, and we ain't seen him since. We gotta call the sheriff and let him know this guy's on the loose!"

Yoo-Hoo!

It was a warm, humid August evening, so humid we could hardly breathe in our upstairs bedroom. The warmth of Lenny lying so close agitated me greatly. I kept trying to push her farther away, but our bed kind of sunk in the middle, and we'd always end up rolling into each other. Mother and Daddy couldn't sleep either. Finally, Daddy gave in.

"Hey, you gals want to sleep out on the lawn tonight?"

Lenny and I jumped out of bed immediately and dragged our blankets and pillows outside, finding the softest spot on the lawn behind our house. Mother and Daddy followed us and made a bed beside ours.

The sky was clear as we lay there, searching for constellations and the Milky Way. We could hear the crickets chirping, and an owl "hoo, hoo, hoo-hooing" off in the distance. We could see huge moths, big as hummingbirds, hovering around Mother's petunias and sipping their nectar.

The petunias smelled fragrant in the evening air, and there was a slight breeze that carried with it all the smells of a farm—straw and hay, animals and barnyards. I loved sleeping outside. It was kind of like camping, a bit adventurous. We seldom had mosquitoes to contend with, because we lived far from any streams or marshes, so it wasn't uncomfortable.

Mother and Daddy were lying on their backs, talking quietly with each other after a long, hard day of work. Lenny was starting to drift off to sleep, and my eyes felt heavy as the blissfulness of sleep crept up on me, too. All of a sudden, a human voice came calling to us through the dark of night.

"Yooooo-hooooo!"

Daddy just rolled over and groaned for a moment, then sat up straight and rubbed his face with both work-worn hands.

"Yooooo-hooooo!" the voice called, again.

"Now what could the Keltys want at this time of night?" Daddy complained wearily. "I sure wish they'd get a phone."

The Keltys were one of the last families who didn't have a phone in our area. Maybe they just couldn't afford to buy one. Times were hard, and money was scarce, so when they needed us, or there was an emergency, they just "yoo-hooed" to us, and vice-versa. We could see their house from our yard, and were obviously within yelling distance.

"Yooooo-hoooooo!" the yodeler called from across the fields, more loudly this time.

Daddy stood up and cupped his hands around his mouth, arched his neck to face the sky and "yoooo-hoooooed" back, letting whoever it was know we had heard their call. Daddy pulled on his pants and tucked in his shirt, slipped his bare feet into his boots and began walking toward the Keltys, through the hog pasture. He didn't have to walk far before Gladys came running to meet him. We were close enough to hear some of the conversation.

"Ralph! Ralph, you gotta come quick! Poor Georgie's in an awful fix!"

"Okay. Slow down, Gladys, and tell me if I need to get the car. Is Georgie hurt? Does he need to see a doctor?"

"Yes, he needs a doctor right away," Gladys answered, and Daddy hurried off to get our newly purchased car, an old 1928 Model A Ford. Gladys came over to see Mother and explain the whole situation.

"Oh, Margie, we'll be forever in your debt if Ralph would see to taking Georgie to the doctor."

"Gladys, what's happened?" Mother asked, with concern.

"Aw, Lordy. Georgie's been seein' a gal. He's really taken with her, Margie, and she's just the sweetest thing! But her ex has been in jail, and he musta got out somehow. Maybe he was paroled, I don't know. Anyway, he found out about Georgie."

"Oh, my," was all Mother could say, her eyes as big as saucers while she listened to the story. Lenny was hanging on to Mother's robe and I was standing close by, glued to the conversation.

"Well, as I said," Gladys continued, "Mamie's man had been in jail for who knows how long. And for what, I cain't recall, but somehow he got outta that jail. And when Georgie came home tonight after seein' her, we heard this hollerin' and yellin' out in front of the house! I ran to the door just in time to see that jailbird haul off and clock Georgie with a tire iron, right between the eyes! Poor Georgie didn't get a chance even to defend himself. He was bleedin' like crazy. He's got an awful gash on his forehead."

Mother gasped, and I sucked in my breath. Wow! This was definitely the most horrible story I'd ever heard! I felt sorry for Georgie. He was a good man who did some work for Daddy once in awhile, but I couldn't wait to hear the rest.

"Well, I screamed for Mama and Pa to come right away, when I seen Georgie fall to the ground like a sack of potatoes."

"And what about the convict? Where did he go?" Mother asked, glancing toward the fields.

I looked, too, but all I could see were shadows of the barn and silo, and the other outbuildings.

"Him? He musta hightailed it into the woods, and we ain't seen him since. We gotta call the sheriff and let him know this guy's on the loose! Or maybe they already know."

I could see that Daddy was almost ready to go. I quickly slipped inside the house to pull on a pair of overalls, then ran back outside just as Daddy drove up with the car. The Model A had a bench seat in front where three folks could sit, and a "turtle" in the back where two more would fit.

"'Bye, Margie. Sorry for troubling you this time of night," Gladys said, climbing into the car.

"Mother, I'm going in case they need help with something," I said, as I ran and hopped into the turtle, before she could say a word.

Daddy turned around, asking, "Is this okay with your Mother?"

"Well, she didn't say no," I said, telling half the truth. I wasn't about to miss all the excitement.

Gladys was so nervous, she talked non-stop for the whole two miles it took to get to the Kelty's place. Taking the roads was kind of a roundabout way to get there. We had to go out our driveway, take a right onto the main road, take the main road all the way around to their driveway, then down to the house. But as the crow flies, it's less than an eighth of a mile to their house.

We pulled into the yard, right up to the Kelty's house, which consisted of a kitchen, parlor, living room, a lean-to bedroom with just enough room for two beds, and an upstairs over the living room. Just inside the kitchen was a cellar door in the floor. I was always afraid to walk on it for fear it would fall in. Truth was, if it could stand up to the strain of holding a big gal like Janie, I'm sure it would have held me just fine. But still I feared falling through, and usually side-stepped the cellar door. Janie came running from the house.

"Oh, Ralph! Son, you're an angel sent from Heaven," Janie cried, as Daddy, Gladys and I got out of the car and headed for the house.

"Where's Georgie?" Daddy asked, as we walked in.

"He's in here. We dragged him to the sofa," Old George said, leading Daddy to the parlor. I followed along behind, being careful not to get in anyone's way.

The parlor was sparsely furnished with a few chairs, a table with a Gamble store radio on it, just like ours, some photographs of Janie's brother who was killed in World War I, and an old pump organ. I'd been told that Nellie played it somewhat, but had never seen her perform.

I glanced at all the calendars hanging on the walls. At the end of the year, when local merchants were handing out free calendars, the Keltys would go from the bank to the drug store, to the feed store, the grocery store, and on, collecting as many calendars as they could. They'd hang them on their walls, all over the house. The house may have been old and sparsely furnished, but it was definitely not drab. With all those colorful calendars on the walls, I'm sure the Keltys always knew what day it was.

"He's out cold, Ralph. And look at that gash on his forehead. Lordy, that's a nasty cut!" Janie cried, folding her hands tightly against her bosom, shaking her head in dismay. She walked over by Georgie and knelt down next to him. She rubbed his shiny, bald head with her hand, so gently.

He looked helpless lying there, his tall frame limp and lifeless. I could tell Daddy was very concerned.

Daddy and Old George talked quietly between the two of them, and then Daddy walked over and slid his arms under young

Georgie's, while Old George grabbed Georgie's ankles and they carried him out to the car.

"Janie, get some blankets and pillows," Daddy ordered, as they hauled the limp man outside. Janie ran back inside the house and came out with a handmade quilt and a feather pillow.

"Mama, you go with Georgie. I'll stay here with Masel," Gladys offered. Janie gratefully accepted her offer, and climbed up into the car with her son. Old George, rather frail, stayed behind with Gladys and Nellie.

"Gladys, you'll need to go back to our house and call the hospital," Daddy said. "Have them call Dr. Spiegelburg. Make sure they know it's an emergency."

Gladys just nodded her head and waved us on. I waved back, as Daddy turned the car around and headed back out the driveway.

Driving off into the night, I scanned the dark woods with eagle eyes, wondering where the convict had gone. Was he hiding out there? Would he come back for Gladys? I hoped not. Evidently, he'd only had a beef with Georgie, so I was pretty certain Gladys and Masel would be safe.

I could feel something rolling around on the floor under my feet. I reached down and picked it up, holding it right before my eyes, trying to make out what it was. It was the broken, wood handle of a pitchfork. I wrapped my fingers around it, feeling a little safer with this weapon in my hand, should the convict decide to ambush our car on the way to Boscobel.

Everyone was quiet as the car rumbled along on the gravel road, Janie and Georgie up front with Daddy, and I in the back. We could hear the crickets' discordant chirps in the tall grass as we raced on by. I could see a mass of spilled stars up in the sky, and the full moon to round things out calmed us all down a bit. It took awhile to drive the six miles into town, but we made good time and arrived at the hospital around eleven p.m.

The hospital was actually converted from an old mansion. Daddy drove the car up as close as he could get to the front door, jumped down and helped Janie carry Georgie into the hospital.

"Leona, run on up to the front desk and let them know we're bringing Georgie in," Daddy ordered.

I got down from the car and leaped up the steps to the

hospital. Dr. Spiegelburg was already there to greet us. The doctor had an office above the Parker-Hildebrand department store in town, and during normal business hours that's where we'd go if we needed to see him. But tonight was different.

"Ralph, what's happened?"

"Well Doc, Georgie got clocked between the eyes with a tire iron and he's out cold," Daddy started to explain.

"A tire iron? What on earth..."

"It's a long story. Maybe Janie can explain it to you, if she wishes."

Dr. Spiegelburg walked with Janie, while a nurse wheeled young Georgie away on a cart. They all disappeared down the hallway.

Daddy and I waited and waited. Finally, the two of them came back out to the waiting room.

"Well, Georgie has a concussion, and we've stitched him up," the Doc said.

"No doubt, he took an awful blow to the head," Daddy agreed.

"Yes, well, we'll keep him overnight just to keep an eye on him, and if he's okay in the morning, you can come back and pick him up." Suddenly the doctor turned toward me and asked, "Say, weren't you the one that dropped a shoe on my head when I came out to check on your Daddy's Uncle about two years ago?"

"Oh, no sir. That was my little sister."

The truth was, Uncle Pete had been living with us for a while back then. He was an old bachelor whose only claim to fame was that he had harvested wheat in the Dakotas, and this made him a world traveler in our eyes. But while he was with us, Uncle Pete had gotten sick, and Dr. Spiegelburg was called to come out from Boscobel.

Lenny and I had been upstairs while the doctor checked on Uncle Pete. In summer, the wood heater and stovepipe were removed from the room Uncle Pete was staying in, leaving a hole in the floor. Lenny had been peeking down on the scene below—actually, we both were—but it was Lenny who accidentally dropped a shoe on Dr. Spiegelburg's head.

We were mortified, and Lenny had gotten in trouble, for sure. So for good measure I reiterated, "No sir, wasn't me, it was my naughty little sister."

Dr. Spiegelburg just laughed.

"Well, Doc, we'll give you a call in the morning, then," Daddy said, then stood up and walked Janie out to the car.

He helped Janie up into the car, then climbed in himself, and we left for home. I think we were all silently praying that Georgie would be all right. It was real quiet all the way home.

Mother and Lenny were waiting up in the kitchen. After explaining everything, we all went to bed.

But sleep would not come easily, for thoughts of Georgie's attacker still lurked in my mind. Where was he now? Was he still hiding out in the woods? Would he attack again? And who would be next? I resolved not to go into the woods until I knew for sure this guy was locked behind bars once again. Daddy did say that the sheriff would be looking for him and that they'd most likely catch him and throw him back in jail. But until I knew for sure, I'd just keep a cautious eye out at all times, and maybe carry that old broken pitchfork handle for a while.

Not too long after Georgie's run-in with the convict, we heard one of the Keltys "yoo-hooing" from across the field again, but this time it was my friend, Masel. Her appendix was about to burst and she was in awful pain.

Daddy and I made the same circuit by way of the roads to get to their house. Daddy gingerly carried Masel to the car and kind of laid her down across her mother's lap, covering her with blankets. I sat next to Gladys and held Masel's legs. During the long ride into town, Masel was becoming a little feverish, and started mumbling something about, "Don't, don't let him get me. Please, please don't let him take me."

"Masel, it's me, Leona. No one's going to take you, you're just dreaming."

I looked up at Gladys. She was holding her daughter's hand, lifting it up to her face and just holding it there.

I knew what Masel was dreaming about, and I felt bad for her. She was always scared that her father would someday come and take her away from her mother. Gladys had been married at one time, and that's when Masel had been born, but the marriage didn't last. Gladys had separated from her husband years before, and we had never known him. That's why Gladys and Masel

A young Masel poses with her mother, Gladys, in this shot from 1937. Masel always harbored secret fears that her father would return to take her away from her mother, and when she was in a delirium during an acute attack of appendicitis, she cried out as though she were dreaming that this nightmare was coming true. Leona soothed her friend on the way to the hospital in Ralph Ahnen's Model A Ford.

were living with her folks. Gladys probably worked out as a hired girl occasionally, to help pay for their keep. Even though Gladys' and Masel's last name was Woodard, to us they were all Keltys.

We heard that Masel's father lived somewhere in Iowa. But this distance didn't ease Masel's fear of being "kidnapped." And now, in these moments of agony, Masel's worst fear was haunting her. There was nothing we could do, except let her know she was safe and that we were going to get help.

In the end, it was better than one of those radio soap operas.

Dr. Spiegelburg came through again: Masel's operation was a success, and she went home to recuperate. By then, the Keltys just adored Daddy and had developed the highest respect for him. There wasn't anything they wouldn't do for him, or Mother, either.

The convict who'd attacked Georgie did end up back in jail, and eventually, Georgie married Mamie.

The Yellow School, the way it looked when Leona, Lenny, Masel and all their friends and cousins went to school there. Leona took the photo around 1937. The school had been built before her father went to school, because he also attended classes there, himself. The bell looks as though it might be calling students to class. Note the pile of split wood waiting for the boys to stack it in the woodshed (background) for use in the wood stove, which heated the one-room schoolhouse.

"Well class, I think we'll get along just fine. All you have to do is follow my rules and things will run smoothly. And remember, these rules are for the good of the whole class, not just for me."

Once we got home from school, Lenny and I listened to our radio programs for an hour, then I went outside. I got the cows from the pasture, fed the chickens, gathered the eggs, and brought down hay and silage for the cows and horses. After supper, we did the milking. My thoughts, though, were not on chores but on what was going to happen the very next day.

Mutiny At The Yellow School

"Come on girls, finish up. Masel will be here soon," Mother reminded us.

Breakfast of fresh "cackleberries," fried pork, homemade bread with lots of butter and jam on it, and fresh milk from the cow sure hits the spot in the morning, before going to school. We heard a tap on the door.

"Mornin', folks," Gladys said, as she walked in the kitchen door.

Masel, right behind her, waved a short, pert hello. She walked with Lenny and me to school every day, and she was always on time. We tried to get to school by eight-thirty, so that we could have half an hour for playing before classes started at nine o'clock. Gladys liked to stay awhile after we left, to sit and gab with Mother about the weather and the day's happenings.

I got up from the table as I ate my last bite of bread, and washed it down with a gulp of milk. I grabbed my lunchbox off the table, my coat from the hook, and headed out the door with Lenny and Masel.

Mother had probably put the usual fare in my lunch: sandwich, apple, piece of cake, and a thermos of hot cocoa. Masel's lunch was usually better than mine, I thought. Gladys would use store-bought lunchmeat for Masel's sandwiches, which we thought was a real treat. A few times, I think she even used some "boughten bread" from the store. I tried this bread once, but didn't care too much for it.

Once out of range of the adults, I asked, "Masel, did you get those 'rithmetic problems Miss Olsen gave us to do last Friday?"

"Yeah, I think so."

"What did you get for an answer to number five?"

Masel reached into her bag and pulled out her tablet.

"Number five? Oh, I guess I didn't do that one. I was gonna ask you."

"Well, I hope Miss Olsen decides to go over these in class with everyone." And on went the chatter, about this person and that farmer, this writing composition and those math problems.

Looking ahead, we could see the yellow schoolhouse sitting on top of the ridge, with its cement steps and platform extending off the front of the building. It was half a mile from my house to the school, but it was uphill all the way and took us nearly twenty minutes to get there. Everyone walked to school, even the teacher, who often boarded with a family in the community.

The teacher couldn't be married, or else she wouldn't be allowed to teach. Miss Olsen was the only teacher we ever had that drove a car, and she also lived with her mother and brother, instead of boarding with a local family. But of all the teachers we had, she was our favorite, hands down. She understood us kids, sometimes better than our own parents did.

One time, Mother was coming to school for her annual visit, just to observe what was going on. I was so excited about this. Miss Olsen told me to write a short essay on anything I wanted, and I was to read it in front of the class when Mother came.

When Mother arrived and it was time for me to give my speech, I suddenly realized that in all of my excitement I hadn't written down a single word! I went forward with my paper anyway, and rattled off a speech worthy of an "A," pretending to be reading it from the paper. I didn't fool Miss Olsen any. She could see what was going on, but I don't think Mother ever knew. Much to Miss Olsen's credit, she didn't say anything, and never asked to see my paper. She was the best teacher a student could ask for.

"Say, I heard what Miss Olsen makes every month for teaching. Guess how much?" I said to Lenny and Masel.

"I'll bet it's more than ten bucks a month!" Masel replied.

"Nope, guess again."

"Twenty dollars a month?" Lenny asked.

"Nope, it's thirty dollars a month! Can you imagine having that much money every single month? I was figuring last night that if Miss Olsen saved everything she made she could save two hundred forty dollars for the whole school year! Can you imagine that? What would you do with two hundred forty dollars, Masel?"

"First, I'd get me a new pair of shoes," she said decisively.

I looked down at Masel's feet as she walked, and the sole of one shoe was coming off. Masel had put a rubber canning jar ring over her toes to keep the sole from flopping. Her other shoe didn't look too good, either. Yeah, she could use some new ones, all right.

"But that's only two dollars. What would you do with the two hundred thirty-eight left over?" I asked.

"Don't know right off. Guess I'll have to think on that one."

"Know what I'd get?" Lenny piped up.

"What?" Masel and I said together.

"I'd get a bunch of new dresses, ones that fit me perfectly."

"What'd you want to buy dresses for?" I asked with disgust. "I'd get a new horse for thirty-five dollars, a new saddle and bridle for eight dollars, and put the rest in the bank."

Thoughts of having money to spend on anything we wanted, at a time when the whole country was in a terrible depression, seemed an unattainable dream. The unemployment rate was almost twenty-five percent, and we were lucky just to have food on the table. Living on a farm at least somewhat guaranteed that we'd have vegetables, eggs, meat and milk.

We were almost to the school, and could see smoke coming out of the chimney. Miss Olsen had already gotten the wood heater going. During the winter, if it was extremely cold outside, the school never got warm and we'd all end up sitting snuggled close to the stove, all day long. But this time of year, in the fall, it wasn't a problem.

We climbed the steps and opened the door into the one-room schoolhouse. We took our coats off and hung them on hooks in the back. The one big room had a large teacher's desk up front, a blackboard on the front wall, a large eight-day clock on the wall, and two rows of desks. The three of us took our seats.

Lenny was way up in the front because she was so young, and Masel and I sat across from each other, four seats back from the front. Masel and I were the same age; in fact, we had been born just four days apart.

The sound of feet stomping up the steps and into the school, and clopping across the wood floor, which still smelled of the sweeping compound it had been sprinkled with prior to being swept the night before; the odor of chalk; the smell of an apple wafting from a lunch pail; children's voices chattering; the wood heater crackling with heat; farm kids coming to school directly from doing chores, in the same clothes worn day after day, most kids only having one bath a week; these were the sights, sounds and smells of the Yellow School.

In the winter, the smell of drying wool snow pants and mittens would add to the aromas. The sweeping compound was really just sawdust mixed with oil and a cleaner or disinfectant. It came in five-gallon cans and was used to clean the floor rather than scrubbing with soap and water. I often wondered if the teacher noticed all this. Miss Olsen didn't seem to mind any, but I thought maybe if we got a city teacher, she'd have to hold her nose all the time.

Miss Olsen brought the class to order, as she stood at the front of the room.

"Class, I have some news to tell you today. I am getting married in a few weeks!"

We all gasped.

"No! Why? Who's going to take your place? You can't leave!"

Miss Olsen tried to shush us by pushing her hands down in the air, indicating for us to pipe down.

"Miss Olsen, you can't get married, 'cuz then you'll have to leave us," Masel said rapidly as she stood up, obviously upset, just as we all were.

"I know. I know this might be kind of a shock, but there's a new teacher coming, and I'm sure you'll like her just as much," Miss Olsen tried to reassure us.

There was an undercurrent of whispers among the students as Miss Olsen tried to go on.

"You'll continue on with the same things we've been studying and nothing much will change around here except for the teacher. Things will just go on as usual."

All of a sudden, we heard the school door slam, and in walked Gus and Virgil Bachim. These two brothers were distant relatives of ours, although we didn't announce this from the barn roof. They had to walk at least two miles to school and were often late.

Miss Olsen didn't miss a beat.

"Oh, good. Gussie and Virgil, it's your turn to get the water this week. Would you please go now?" Miss Olsen asked politely.

Gus bent his neck back, dropped his shoulders and rolled his eyes, but Virgil grabbed the buckets by the door and back outside they went. They had to walk down the hill, through the sheep pasture, across Billy Martin's field to the windmill pump, then carry the water back to school.

Half an hour later, the Bachim brothers came back and poured the water into a large crock with a spigot, which sat on a shelf at the back of the school, next to the washbasin. The communal dipper hung on the wall. Suddenly, we all heard something splash on the floor.

"Aw, shucks!" we heard Gus whisper hoarsely to his brother.

"What's the matter, Gussie?" Miss Olsen inquired.

"Aw, I spilt some on the floor."

"Well, no matter, just sweep it out the door."

So Gus proceeded to sweep the water toward the door. We had our backs to Gus, but could hear the swishing of the broom.

"Miss Olsen? This ain't working. The water's not makin' it to the door."

Everyone giggled a little as Miss Olsen got up from her desk and went back to investigate.

"Oh, I see."

"It's just spreadin' all over the place, and I can't sweep it out the door. But I tried my best, Ma'am."

"Gussie," Miss Olsen said, affectionately tousling his hair, "Just go back to your seat. It'll dry in no time."

Miss Olsen went back to her desk while Gus and Virgil took their seats in the last row, right in front of the school's "library." Our library consisted of a cupboard with books arranged from the lowest grade on the lowest shelf, up to eighth grade on the top shelf. By the time I was in fourth grade, I had read nearly every book in the cupboard.

"As I was saying," Miss Olsen continued, "Things will go on as normal. The new teacher will be here soon. Unfortunately, today is my last day, because I need time to make the wedding preparations, and I..."

Miss Olsen had no time to finish her speech before the school door opened again. This time a young woman, maybe eighteen years old and pleasingly plump with brown hair and brown eyes, smartly dressed and very pretty came walking into the classroom.

Poor gal, she didn't stand a chance, right from day one.

"Oh, here she is already," Miss Olsen said, rather quietly.

The strange woman walked up to the front of the classroom, right up to Miss Olsen.

"Miss Olsen, I presume?" the woman asked.

"I am, and you must be the new teacher," Miss Olsen said, trying to look nice and talk pleasantly.

"Yes, I'm Miss Rubin."

"Okay. Well, students," Miss Olsen said, turning toward the class, "This is Miss Rubin, your new teacher." She gestured for us to say hello, but no one responded. Finally Miss Olsen said out loud, "Class, please welcome Miss Rubin."

"Hi, Miss..."

"Hello teach..."

"Miss Ru..." The greetings trickled in. It certainly wasn't a hearty welcome, but this didn't seem to bother Miss Rubin.

"Class, why don't you take out your readers while I explain a few things to Miss Rubin," Miss Olsen said.

We pulled out our readers and pretended to read while the two teachers were busy discussing things behind Miss Olsen's desk, or, what was now Miss Rubin's desk. Oh, this was not going to be fun. Miss Rubin probably couldn't crack a smile if her life depended on it, she seemed so stiff and uncomfortable.

I watched the two teachers up front, Miss Olsen showing Miss Rubin some books, explaining things in her pleasing, familiar voice.

Miss Rubin didn't say much, just nodded her head and said, "Mmmm hmmm..."

What does that mean anyway? Does it mean, "Yes?" Does it mean, "I understand?" Does it mean, "What?"

I was so disturbed by this turn of events, I was fuming. Thoughts of running away crowded my mind. I couldn't stand to see Miss Olsen leave. My mind was whirling with bad thoughts about how we could get rid of Miss Rubin.

Good heavens, Leona, give the lady a chance, at least, I thought. "Okay, okay. I'll give her ONE chance. At least she's pretty, and she looks really young, fresh out of school. Hmmm, no experience... this might be more fun than we realize."

Half an hour later, Miss Olsen said, "Okay, that's that," walked away from her desk, and came towards us, reaching out with open arms.

"Goodbye, sweet children. I'll miss you all and I'll be thinking of you."

We all jumped out of our seats—except for the older boys, of course—and surrounded Miss Olsen with hugs and kisses and endearments of all kinds. I could see that Masel had tears in her eyes. She was very attached to Miss Olsen.

Our dear teacher finally had to tear herself away from us, wiping a tear from her cheek. Miss Olsen picked up her personal belongings and walked out the door. We all just stood staring after her, wishing she'd come back, hoping this was all a mistake. Abruptly, we were brought back to reality.

"Students, sit down," Miss Rubin said, trying politely to get our attention. When that didn't work, she said more forcefully, "Children, sit down! Now I am a rather strict teacher. I'll tell you this right off the bat, so don't you get any notions in your heads of being able to goof off, take advantage of the new teacher, or anything of the kind," she began.

My cousin, Bernice, piped up, "Hey, is this your first time teaching? You look awful young."

Bernice was a free spirit, and didn't care too much about what folks thought of her. She had a fondness for mischief, and was definitely one tough cookie, like time she broke her arm and set it herself.

Bernice's younger sister, Beatrice, was a little more subdued. She was born with a dislocated hip, and as a baby she was in a cast from her hips down, and could only lay on her back. She was a sweet, patient little girl who never complained, unlike her older sister. After her cast came off, she still had a hard time walking and running. She seemed to do pretty well in school, though.

"We're doomed," I mouthed to Masel, across the aisle.

"Now, that is one thing I will not tolerate in my class room—talking out of turn. If you want to say something, raise your hand. If you want to get up to get something, raise your hand. If you need to use the outhouse, raise your hand. If you need a drink of water, raise your hand. If you need to talk to your neighbor," Miss Rubin said, looking directly at me, "Raise your hand. Do I make myself clear?"

Nobody said a word.

"Do I make myself clear?" Miss Rubin repeated in a louder voice.

"Yes," we all answered quietly.

"Good. Then it's time to get to work."

The rest of the day was spent with our noses in books and paying strict attention to the teacher. Everyone was so quiet, probably still in shock. Once, in the afternoon, Masel got up to go to the outhouse without asking. She just got up and started heading for the door, like we always had before.

"Where are you going, young lady?" Miss Rubin asked boldly.

"Oh, I forgot, Ma'am. Can I go outside?" Masel spit her words out as quickly and quietly as she could.

"For what? And please slow down when you speak. You need to speak clearly, pro-nun-ci-ate your words, Miss… What's your name?"

"Masel. Masel Woodard, Ma'am."

"Please don't call me ma'am. Call me Miss Rubin. And no, you may not go outside. You can wait until recess time, which will be in half an hour."

With that, Miss Rubin continued writing on the blackboard behind her desk at the front of the schoolroom.

I could always tell when Masel was nervous. If she started using "Ma'am" and "Sir," this was a dead giveaway. I looked up at Masel, who obviously needed to relieve herself, and gave her a knowing look. Masel sat down, crossed her legs as tight as she could, and held out until recess.

When Miss Rubin announced it was recess time, Masel bolted for the door and ran for the girls' outhouse as fast as she could. Some of the older boys were laughing.

I was fuming. How dare she do this to my friend! I was thinking, as I walked out the door. It was lucky Masel didn't decide to just go on the floor. Would have served her right...

I walked outside, over to the girls' outhouse. Masel had been in there awhile.

"Masel, are you all right?"

"Yeah, I'll be out in a minute," was her muffled reply.

When Masel opened the door, I could see her face was flaming red from embarrassment.

"Masel, we gotta do something," I said seriously. "Come on, follow me."

We walked to a private spot at the back of the school. Bernice and Beatrice were already there.

"Leona, we gotta do something. Miss Rubin acts like she's a dictator or something!" Bernice complained loudly.

"I was so upset, Masel, when she made you sit down. I wanted to give her a piece of my mind," Beatrice said in her sweet voice.

Bernice stood up on a chunk of firewood and took charge.

"Listen, we need to make a plan. We can't just let her walk all over us! When Miss Olsen was here, we were on our honor. She trusted us to do what was right, and we did!"

"But this teacher, she ain't trustin' us right off the bat, and we ain't done nothin' yet!" Masel added.

"Okay, guys, it's time we made peace with the boys so we..." Bernice started to say.

"What? What do you mean by 'makin' peace' with them?" Masel whispered hoarsely, pointing over at the group of boys standing by the woods.

"Aw, come on. They're not that bad. Besides, we need their help if we're going to form a plan to get rid of Miss Rubin," I said, sticking up for the boys.

"Well, I guess, seein' as how it's for a good cause an' all," Masel said.

"Speakin' of the devil," Bernice said sarcastically.

I turned around to see Gus, Virgil, and Norman walking toward us.

"Leona, you got somethin' up yer sleeve?" Norman asked.

"Well, we were just talking about you guys. We need to combine forces against Miss Rubin..." I started to say, but was cut off.

"You bet we do! Miss Olsen sure never had any dumb rules like, 'Gotta raise your hand if you want to get a drink,' " Gus mocked in a satirical, falsetto voice.

"Well, if you'll let me finish what I was saying," I scolded. "We need to put our heads together and come up with a plan to totally demoralize Miss Rubin, and..."

"What's demoralize?" a small voice piped up.

I looked around, then saw Lenny hiding behind one of the boys.

"Lenny, you gotta swear not to tell anyone, you hear?" I said.

"I promise. What you guys going to do to Miss Rubin? She's kind of pretty, don't you think?"

"Well, yeah, she's pretty, but that don't mean a thing when she's trying to put all these rules on us like we was little kids or something."

We all sat down in the tall grass, huddled together, sharing ideas, making plans. Never in the history of boys and girls has there ever been a group more unified than we were. Suddenly, we heard the bell ring. Recess was over. We stood up to walk toward the school, but before anyone could take a step, Bernice put her arms out to stop us.

"Remember, don't do anything today. Nothing until tomorrow, and then we'll let 'er have it. Okay?"

Everyone nodded and went back in the school. Bernice was almost giddy with delight.

The last part of the school day went as normal as could be. Everyone raised their hands when they needed something, asked politely, but with a bit of a smirk on their faces, and a few giggles here and there. No one caused a fuss or trouble of any kind, almost as if Miss Olsen were still here. Just before school was out and we had all done our chores, like washing the blackboard, clapping the chalk from the erasers, or bringing in firewood, Miss Rubin stood at the front of the class.

"Well class, I think we'll get along just fine. All you have to do is follow my rules and things will run smoothly. All right, everyone? And remember, these rules are for the good of the whole class, not just for me."

"Yes, Miss Rubin," we all chanted together. I think we all felt a little patronized, to say the least.

"Good! You're dismissed for the day."

Once we got home from school, Lenny and I listened to our radio programs for an hour, then I went outside. I got the cows from the pasture, fed the chickens, gathered the eggs, and brought down hay and silage for the cows and horses. After supper, we did the milking. My thoughts, though, were not on chores but on what was going to happen the very next day.

At breakfast, Mother asked about the new teacher.

"Well, what's the new teacher like? What does she look like? Come on girls, you haven't said much about her."

"Well, I guess she's kind of pretty, if that matters. And she is really strict!"

"Oh? That's good. It'll take a strong woman to keep those boys under control. Gus and Virgil, Norman... they can cause trouble when they have a mind to."

"Mother, what's demoralized mean?" asked Lenny. My eyes shot open immediately, as I gave Lenny a swift kick under the table. Luckily, Mother had her back turned to us.

"Why do you ask about such a word? It means to throw someone or something into confusion, I guess you could say," Mother answered.

I looked sternly at Lenny across the table. She was oblivious.

"Is that what you want to..." Lenny started to ask me. I cut her off quick as a hot knife through soft butter.

"Lenny! Tell Mother what you learned yesterday. You know, your new spelling words?"

Lenny then proceeded to tell Mother her new words while I gulped down the rest of my oatmeal, got up from the table and was out the door, even before Masel had arrived. I waited outside, my heart pounding at the close call.

Oh, if Mother ever found out what we had planned to do, she would have skinned my hide for sure. No, actually, it would have been Daddy, and that was ten times worse! The leather strap was hanging just inside the kitchen door, a strap that knew my behind all too well!

But not Lenny's, of course. She was a good girl; never got into trouble, never did anything wrong, Mother's sweet little girl. Not me. I was the tomboy, the troublemaker, the rebellious one. There were times when I wished I could be more girlish and good like Lenny, but on the other hand, I was always wishing I was a boy, or could at least have a brother to play with. Well, whatever I was, I couldn't let Mother and Daddy find out about the plan to purge our school of Miss Rubin.

Deep in thought, I didn't hear Masel walk up behind me.

"Hey, you! We ain't gonna chicken out today, are we?" Masel said.

I jumped, a bit startled.

"Shhhhh! No. No, of course not. We're all in this together, and it's for the good of our school." We started to walk without Lenny.

"Right, Leona! We cain't just let Miss Rubin come waltzin' in and takin' over everything, like she's the... the... Czar, or something!" Masel said, emphatically.

"The Czar? Where'd you get that from?" I asked incredibly.

"Didn't you read the world history lesson last night?"

"No, we weren't going to, remember? That was part of the plan, Masel!"

"Oh, yeah. I forgot. Well, I'll pretend like I didn't, okay?"

"All right. Let's not talk about this in front of Lenny. She almost gave me away this morning."

"Really? What'd she say?"

"Never mind. Here she comes, the little tattler."

Lenny was running to catch up with us.

"Wait for me, guys!"

Climbing the hill to school, I was nervous. Talking about doing something wicked is one thing, but actually doing it is another.

It's for a good cause, I reminded myself as we walked in the door, hung up our coats and took our seats in the classroom. I looked around.

Miss Rubin sat in front at her desk, her hair done just so, hands folded in front of her, waiting for us to sit down. She wore the same dress again, but this wasn't unusual. We all wore the same clothes most of the week. She started checking her attendance book as all the students entered the building.

As I walked to my seat I was wondering where the boys were, and Bernice.

"Hi, Leona," a familiar voice sounded from behind me.

I turned around and looked up at Kathleen, Norman's sister.

"Hi, Kath. Where were you yesterday?"

"Wasn't feelin' good." Kathleen bent down and whispered in my ear, "Norman filled me in on what's happening today."

"Good," I indicated with a nod of my head, glancing in Miss Rubin's direction. Just then, I happened to see Gus streaking across the backyard, running from the boys' outhouse, with Norman and Virgil in hot pursuit.

The last to enter were the big boys, Gus, Virgil and Norman, and Bernice was tagging along behind. They took their seats. Gus looked a bit winded. Someone poked me in the ribs and I turned around.

"What?" I asked Norman. He just winked and I winked back.

"Good morning, class. Ready to start our second day? Everyone remember the rules?"

Gus stood up without raising his hand and said, "No Ma'am, could you refresh my memory, please?"

Miss Rubin looked a little perturbed, but she rattled off her list of rules from the day before, adding a few more at the end for good measure.

"Gus, I will not tolerate outbursts in this class. You will have to go stand in the corner the next time this happens, understood?"

"Yes, Ma'am" Gus drawled out.

"Don't call me ma'am. I am Miss Rubin and that is what I wish to be called. Understood?"

"Yes, Ma'am, uh, Miss Rubin," Gus said, smiling as he sat down.

Gus and Norman chuckled quietly as Miss Rubin had everyone stand up beside their desks for hand and fingernail inspection. I smiled knowingly.

"Leona, hands, please?" I held out my hands for Miss Rubin to inspect. They were badly stained from picking walnuts recently. "What's this? You're hands are filthy."

"No, Miss Rubin. Don't you know a walnut stain when you see one? From the hulls?" I said, innocently. "You can't say anything about that, you know."

Miss Rubin's forehead creased and she went on to the next student. She had obviously never picked walnuts.

"I hope you older ones read your world history assignment last night," Miss Rubin said, as she examined everyone's hands. She then went up front to her desk.

Gus jumped up from his seat again.

"Did we have an assignment, Ma'am?"

Miss Rubin put her hands on her hips and narrowed her eyes at Gus.

"Gus, that's it. I've told you to call me Miss Rubin and you can't seem to remember that. You also seem to forget my number one rule of raising your hand when you have a question. And yes, you did have a reading assignment last night!"

Miss Rubin was almost yelling toward the end of her long-winded sentence. She pointed her index finger in the direction of the front left corner and looked straight at Gus. He jauntily walked up to the front of the room, smiling, walked right over to his corner, turned around and took a bow. This last act of heroics started all the students laughing, and I could tell the war had begun.

After quieting the classroom down, Miss Rubin tried to pick up where she had left off, asking questions about the history lesson. Gus had his head leaning against the corner, his back to the room. I watched as he pulled a soft, small bag out of his pocket. He had the little bag in front of him and I began to wonder what he was up to.

Miss Rubin had her back to Gus, and as she continued her questions, I watched him slowly bend his knees and squat down in the corner.

"Leona, what is the capital of Russia?" I jerked my gaze back to the front. "Please stand up as you give your answer."

I stood up hesitantly, not answering right away.

"Well, Leona? What is the capital?"

Well, it's now or never, I thought to myself. I took a deep breath and glanced over at Gus again. My eyes about bulged out of my head as I saw a thin, green snake slithering out from the corner, where Gus was now standing again. He was trying to look behind him.

"Leona!" Miss Rubin said sternly.

"Uh, well… I, see…" I stammered.

I looked down at the floor. The snake was slithering under Miss Rubin's desk now.

"To tell the truth, Ma'am, I didn't read it," I said all at once. There, I did it. Now I'm in trouble, I thought.

"Leona, do not call me ma'am, and the capital of Russia is Moscow! Go to the blackboard and write Moscow a hundred times. Maybe this will help you to remember to do your homework next time."

I walked up to the front, skirting wide around Miss Rubin's desk, picked up the chalk and began my punishment. I looked over at Gus in the corner, and we silently laughed at each other. Masel raised her hand.

"Yes, Masel?"

"May I go to the outhouse, Ma'am?"

"Yes, you may, but Masel, please don't call me Ma'am," Miss Rubin said, sounding kind of discouraged.

"Okay, Miss Rubin," Masel said as she turned around to walk outside.

As she passed by the desks, Bernice tried to get Masel's attention without attracting notice, but Masel was oblivious. As Masel walked right on by, Bernice finally gave up and rolled her eyes.

"Bernice, is there a problem?" Miss Rubin inquired.

"No, Ma'am. I'm just fine!"

"Bernice, what was all the motioning about, then?"

"Oh, nothin', Ma'am. Just got an itch in the middle of my back," Bernice said, pretending to reach over her shoulder and scratch.

"Bernice, you just called me Ma'am twice, and I'm tired of having to repeat myself. In the corner, now!" Miss Rubin pointed to another corner.

Bernice got up and nonchalantly walked to her corner. Just behind Miss Rubin's back, Bernice gave everyone a "thumbs up," and made a funny face at the back of the teacher's head. The little ones started to laugh.

Miss Rubin told everyone to quiet down again, and she went back to her desk.

Masel was gone a long time, and finally Miss Rubin sent Kathleen out to check on her. A few minutes later, Kathleen came walking back into the school and up to Miss Rubin's desk. She whispered quietly, "There's no toilet paper in the outhouse."

"Oh? Well, you have permission to go to the woodshed and get some," Miss Rubin whispered back.

Kathleen left again, but minutes later, was back at Miss Rubin's desk.

"Miss Rubin, the toilet paper's all gone. I can't find it."

Miss Rubin then got up from her desk and went to see for herself. While she was gone, the grass snake came out from under the desk, but no one noticed it yet.

Miss Rubin came back in and walked up to the front of the room, hands on her hips, stopped and turned around in front of her desk.

"All right, it seems that someone has stolen all the toilet paper from the woodshed, and I want the guilty one to identify him or herself right now." Silence. "Bernice, what do you know about this?"

"Nothin', Ma'am. I have no idea where the toilet paper went. Maybe Gus used it all up the last time he was in the outhouse," Bernice volunteered. She looked over at her partner in crime in the other corner, smirking.

Gus narrowed his eyes and gave Bernice a rather menacing look. He didn't seem to appreciate Bernice's comment.

"Norman, would you go check in the boys' outhouse, please?"

Norman left and came back in a minute.

"Norman, did you find anything?"

"Yes, Ma'am," he said.

"Don't call me Ma'am, and what did you find?"

"Lots and lots of toilet paper, Miss Rubin," Norman said with a straight face.

"Well, would you please put it all back in the woodshed and give Kathleen a roll to give to Masel?"

"Sorry, Miss Rubin, but I can't."

"Why not, for heaven's sake?"

"There's a note attached to the box that says, 'Anyone who touches this toilet paper is taking the risk of being dunked head-first down the outhouse hole!' "

Miss Rubin looked disgusted.

"Of all the tricks to play, this was the best you could come up with, boys? Kathleen, you have my permission to enter the boys' outhouse and get a roll for Masel. But Kathleen politely refused, stating that it was not proper for a girl to be entering the boys' outhouse. Miss Rubin hastily handed Kathleen some paper, saying, "Well, here then, give this to Masel."

Miss Rubin turned around for a moment, as if to gather her thoughts, then turned back to the class. "Children, this behavior is unacceptable. I cannot..."

"Aieeeeee!!! Miss Rubin!!!" Lenny shrieked.

I turned around, and so did Gus and Bernice. Lenny was pointing to the floor under Miss Rubin's feet. Miss Rubin began to scold Lenny for screaming, but glanced down hesitantly and saw the green garter snake under her feet.

"Aaauuggghhh!!" Miss Rubin screamed, then stopped herself short. She took a quick sidestep away from the snake, as some of the young girls were climbing on top of their desks. Miss Rubin took a deep breath, flared her nostrils and said rather calmly, "Norman, would you come up here please and take this snake outside, where it belongs?"

Norman raised his hand, waiting for Miss Rubin to acknowledge him.

"Norman! What are you raising your hand for?" she asked quickly. "Would you please get up here, now?"

"Well, Ma'am, you see, I'm deathly afraid of snakes..." Norman began his speech.

"Virgil! Come get this thing out of here, now!"

Virgil, too, raised his hand, waiting to be called on.

"Virgil! For Pete's sake, put your hand down and do as I asked!"

Virgil kept his hand in the air.

"All right, you want to play games? Is that what's going on here? Fine, Virgil, put down your hand, I'm calling on you!" Miss Rubin said angrily.

"Ma'am, I'm sorry to say, but I'm allergic to snakes, and if I touch one I'll just break out in sweats and hives like you wouldn't believe," Virgil said, all seriousness.

Kathleen and Beatrice were snickering behind hands, which covered their mouths. Gus was in the corner, holding his sides from laughing so much. Bernice was in her glory, and I was up front, amazed at the participation of everyone to "demoralize" Miss Rubin.

It was working! It was actually working! Miss Rubin walked quickly behind her desk and over to Gus. She grabbed him at the back of his neck and forced him over to where the snake now lay coiled under her desk.

"Gus, get that snake out of here now, or you're not coming back for the rest of the year. Do you understand me?"

Gus smiled his infectious grin, bent down and picked up the snake by the tail. He dangled it in front of Miss Rubin—accidentally, of course. She jumped back and hit the blackboard where I was standing, getting chalk all over the back of her brown dress.

Gus proceeded to walk out the door and let the snake go, calling in a mournful voice, "Goodbye, dear snake. We part with such sweet sorrow..."

The whole class broke into an uproar over Gus's little poem. Miss Rubin grabbed a yardstick from the corner and slapped it on her desk so hard, she broke it in half.

Instantly, there was silence. Everyone stopped in their tracks. Miss Rubin stood behind her desk, resting her hands on the top and leaning forward toward us students.

"Everyone sit down, now!"

Shocked at her tone of voice, everyone immediately obeyed.

"Okay, that's better. Now let me tell you children something. You think you're going to scare me away? You think by acting up, that you'll drive me away? Do you think I'm so easily scared? Well, let me tell you, I'm not! I understand you may be upset that Miss Olsen has left, but that's no call for misbehaving. And I'm here to stay, whether you like it or not. So you might as well get used to me, get used to my rules, and get used to following those rules. Understand?"

"Yes, Miss Rubin," we all chanted in unison.

"Now, we're going to start over. Leona and Bernice, go back to your seats. Gus, back to yours, and we're going back to Square One. Repeat after me: I will raise my hand when I want to speak," Miss Rubin demanded. "Go on, say it, all of you!"

"I will raise my hand when I want to speak."

"I will raise my hand when I want to get something."

"I will raise my hand when I want to get something," we all parroted, as Miss Rubin ran off her list of rules, like we were little toddlers or something.

After school was out that day, Miss Rubin called me back to her desk. Waiting until all the students were gone, she then asked, "Leona, what is going on here? Why are all the students so against me? I'm not mean, am I?" I didn't know what to say at first. "Go on, you can tell me. Tell me the truth. Don't be afraid."

"Well, Miss Olsen has been here for as long as I can remember, and she didn't have any list of rules," I started to say and then stopped.

"Go on."

"Miss Olsen, she just trusted us to behave and to act on our honor, and maybe because she trusted us we just did what was right."

Miss Rubin nodded her head up and down a few times.

"I see..." was all she said, but I noticed her eyes were watery.

"Can I go now?" I asked quietly. Miss Rubin didn't say a word, but silently waved at me to go on home. As I was halfway down the classroom, I stopped and looked back. "Oh, and there's one more thing. Miss Olsen used to play with us at recess and lunchtime," I said, and walked on out the door.

I walked into a waiting crowd.

"What'd she want, Leona? Are you in trouble?"

"What'd she say?"

"Is she really mad?" everyone asked me, all at once.

"No, no. She didn't seem mad, and I'm not in trouble, but I think she was starting to cry."

"Should we continue with our plan tomorrow?" someone asked.

I shrugged my shoulders, looking to Bernice for an answer. Norman ushered everyone away from the steps and down the hill a ways.

"No way are we going to give up, tears or not!" Bernice expounded. "We're just getting started, and unless you want Miss Rubin to stay, we gotta keep the pressure on. Okay, guys? No backing off now."

Everyone agreed, then went home for the night.

The next day was worse than the first. Miss Rubin ran out of corners to put us in, the toilet paper was still held hostage in the boys' outhouse, and she stubbornly stuck with her list of rules, which we stubbornly rebelled against for two whole months.

Poor Miss Rubin, she was in tears almost every day. At times I nearly felt sorry for her. That is, until Christmas.

Our Christmas program was a disaster, because we'd forget our lines and mess up on our poems, "accidentally" knock over props and "misplace" others. It was the worst and funniest Christmas pageant ever. The parents were laughing so hard, they had tears streaming down their faces.

But for Miss Rubin, it was the last straw.

After the program was over, Miss Rubin quietly announced to one of the school board advisors that she was leaving, and that we'd have to find another teacher. She mentioned something about being "incorrigible, unruly, mean-spirited, and unteachable."

The shocked parents didn't know what to say as Miss Rubin left that night, in tears. A couple weeks later, Miss Fry appeared to take Miss Rubin's place.

Saturday night was bath night.

"Leona, go get some water please," Mother would say.

So I'd have to walk out to the black hand pump in front of our house, and pump the handle up and down, up and down, until the water splashed into our metal buckets. I can still hear the creaking and clumping sounds the pump made. I'd have to make several trips into the house and back out again. All the while, Mother would be heating the water on our wood stove in a huge copper boiler. My arms would get pretty tired.

We would eat supper while the water was heating, and after supper was done, we'd start our bath routine. Daddy always liked to remind us, "Remember now, girls, wash up as far as possible, wash down as far as possible, then..."

"Wash possible!" we'd chime in, completing our family joke.

Lessons From Grandma

Lenny always got to go first because she was the youngest. Mother would put the washtub in front of the wood stove and leave the oven door open, so we wouldn't catch a chill coming out of the tub. Catching a chill was always a worry to Mother, particularly in the springtime. She'd always remind us, "Don't sit on the ground now, or you'll catch a chill!"

Lenny could sit down inside the tub completely, and Mother would help wash her from head to foot, scrubbing and scouring her scalp with soap.

The scrubbing part wasn't so bad. It was the final rinse that was such a shocker. Mother collected rainwater from the house roof in a big stone crock. She'd pour this cold water over our wet heads to rinse out the soap, but more importantly, to close the pores in our skin so we wouldn't catch cold! I never understood Mother's logic when it came to dousing us with cold water to avoid catching colds, but Mother had her ways of doing things, and who was I to tell her differently?

Once Lenny's bath was over with, it was my turn. Because I was too big to sit down in the tub, I had to stand and could only take a "spit bath." Mother held a blanket up around the tub as best she could for privacy, and allowed me to wash myself. Once

Jacob and Veronica Wetter, Leona's maternal grandparents are probably in their early twenties in this photograph. They were married in 1896, so it is possible that this is their wedding portrait, though if so, it would differ from most traditional poses of the time, since Jacob is not sitting down. They are wearing what were most likely their only suits of formal clothing, which did multi-purpose duty for occasions ranging from church services and funerals to weddings and important business meetings, such as those with a banker. This portrait was probably made at the Anchutz Photo Shop in Boscobel, as it was the only such studio for miles around, even when Leona was growing up.

"possible" was washed, and the "shocker rinse" had been administered, I dried off quickly and slipped into my pajamas, glad that bath night was over for the week.

The next day would be Sunday, and another difficult task lay ahead: getting dressed up for church.

When I was about ten years old and Lenny was five, Mother and Daddy decided it was important for us to go to church. So, every Sunday we'd head into Boscobel to the Methodist Church. I don't know about Lenny, but I wasn't too happy about this decision, mostly because I had to wear a dress and act like the other girls.

Reverend Taylor was the pastor and Mother loved him as she loved her own father. Daddy wasn't quite as religious as Mother, but he did go to church to please her, and to set a good example for us girls. Eventually, he would come to realize the importance of God in his life, and none too soon, but for now, Daddy pretty much just went along with Mother.

After the main service, Lenny and I usually had to go to Sunday school. It was a free-for-all. The teacher had no control over the kids in our class. Granted, she had some troublesome boys, but I never did learn anything there.

Once in a great while, we were excused from church. Daddy would give us each a nickel so we could walk down town to Archie Shaw's and buy two ice cream cones. Archie was always open on Sunday, his little store sitting right next to the municipal building.

Most Sundays, though, we were made to sit in church. To me, the sermons seemed long and drawn-out, and my attention wavered. Sitting in an old building, stuffed like sardines into old pews, listening to the same songs over and over again, my ears hurting when some older ladies' voices cracked on the high notes, just wasn't my thing. I liked seeing my friends, but I was always relieved when the sermon was over and we could get outside.

I guess I was a little rebellious and thought too much religion would cramp my style, and make me have to stop doing some of the things I liked best. I wasn't repentant at all, much to Mother's dismay. I really didn't think I had anything to repent for! In my mind, I was doing fine just the way I was.

Oh, how I preferred the fresh air and wide open spaces, and I'd daydream of riding horses while I should have been listening to the sermon. Maybe I kind of innately knew that God was out there too, in the fields where I'd ride, and especially at my secret places. I just couldn't imagine why God would want to be in a stuffy old church every Sunday when He could be anywhere else.

One Sunday after church, we went out to Grandpa Wetter's farm. Lenny and I climbed into the back of the car, and Mother and Daddy sat up front as we started off down the dusty road.

From our place, we could look West about ten miles and see a huge old oak tree off in the distance. This marked the spot where Grandpa Wetter's farm was, with a knob of white pine trees on the next hill. The farm was nestled in a valley at the bottom of a very steep hill.

There were several of these pine groves in Marion township, with steep cliffs that had trees growing right out of the side of the cliff. The "pine knob" was the place to play when all of our cousins were together at Grandpa Wetter's farm. No one ever got hurt there, despite the slippery pine needles and sheer drops of seventy-five feet or more.

This particular Sunday, Daddy was feeling kind of low. His favorite workhorse, Old Ted, had died a few days before. He had suddenly taken ill and we never knew what it was that did our old faithful horse in, except maybe old age. Normally, when an animal died on our farm, it was thrown in the ditch by the hollow, but Daddy just couldn't do that to Old Ted. He loved that horse too much to just discard his body in a ditch.

So, Daddy dug a huge hole in the ground out behind the barn, and there laid to rest one of the best horses he ever had. I never knew Daddy to cry much, but when he buried Old Ted, there were tears running down his weathered face.

As I stood next to Daddy during Ted's burial, we talked of all the fun times we'd had with Ted. I remembered coming up with the idea of riding Old Ted all the way to Grandpa Wetter's farm. The folks had said it was okay. I had been about ten years old, and had no saddle or fancy bridle, so I rode bareback.

I left early on a Thursday morning in November during a

short school vacation. I had reached Uncle Al's first, and Aunt Nellie called Mother to give her a progress report. Then I went on to Grandpa's. Aunt Gert was still living with her parents there then, and I spent a couple days with her.

"Yeah, you were pretty proud of yourself, weren't you?" Daddy asked, leaning on his shovel and wiping a tear from his face.

"Yes, I was, but it didn't last long. It turned cold and rainy the day I was going to ride for home, remember? And Grandpa was worried about me riding home in the cold drizzle."

"Actually, he didn't approve of your journey in the first place."

"I know, but I still think he over-reacted when he and Uncle Pat decided to bring me home in the buggy, with Old Ted trotting along behind. I think Ted was a might disappointed, as well!"

"Maybe so. He was a fine, sensitive horse. I'm sure he sensed your dismay."

"They made me sit between them until we met you on Dry Hollow road, and..."

"And I brought you home the rest of the way," Daddy finished.

"Mmmm, hmmm. I was so humiliated, not to have finished my trip by myself. And cold!"

"But Grandpa and Uncle Pat were only looking out for you. Maybe someday you'll get a chance to redeem yourself. But this time, don't go in November, go in July."

Daddy and I had laughed together, then finished our task of burying Old Ted.

As our car pulled in by the barn, Aunt Gert came walking out of the house to greet us.

"You made it! Are you hungry, girls? Come, let's get something to eat before you dry up and blow away like a dandelion seed on a windy day."

We liked Aunt Gert. She was kind of quiet, friendly, and always pleasant. Fairly thin, of medium height, and pretty, Aunt Gert had wide-set eyes and a good amount of beautiful brown hair on her head. But of all the things Aunt Gert was, mostly she was frugal. She was, hands down, the best person for making a dollar stretch and using whatever she had on hand to make do for her needs. She also helped to take good care of Grandma Wetter.

When Grandma first "lost her mind," Grandpa put her in

Another photo from 1937, taken by their mother, shows the Ahnen girls posing in their dresses (an unusual occasion for Leona) before attending church and Sunday school at the Methodist church in Boscobel. Ralph had been raised a Catholic, but Leona believes that an incident in his childhood, in which his family was turned down for some much-needed financial help by their church, caused him to become a Methodist when he married Margie Wetter. In the background of this shot you can see the granary and shed on the Ahnen farm.

the asylum at Mendota in Madison, but by and by, he didn't think they were taking proper care of her and brought her home again. At first, Mother helped Grandpa take care of Grandma, and then Aunt Gert and Uncle Pat took over when Mother left at the age of twenty-six to get married. Gert and Pat were sixteen and ten years old, respectively, at the time, and inherited full responsibility for keeping the house, baking, washing, cooking and making the best pies.

Aunt Gert and all of her brothers and sisters were allowed to go to school through the eighth grade, but no further. Grandpa Wetter was adamant that none of his children would go to high school. He knew deep down that if they left to go into town for high school, they'd never come back to the farm, and he needed them there. Besides, it was rather unusual in those days for anyone to go on to high school. For most people, it just wasn't considered necessary.

We could smell pies baking before we even reached the front door. Aunt Gert still made pies, and had been baking apple and cherry ones when we arrived.

"Sit down, girls. You, too, Maggie," she quietly ushered, pulling out chairs for us to sit down in. Mother's family members were the only folks who called her Maggie instead of Margie.

Aunt Gert pulled a pie out of the wood stove oven, piping hot and ready to melt in our mouths. "We'll let it sit for a minute, to cool down so it don't burn your mouths. How about a glass of milk to start with?" she asked, looking at us.

Lenny and I both nodded our heads As we sat eating our pie and drinking the cold milk, we could hear some thumping or knocking sounds coming from upstairs. Grandma's voice came drifting down the steps, a gentle, older voice speaking in German.

"How's Grandma today?" Mother asked Aunt Gert.

"Oh, she's fine, but didn't eat much breakfast. I do wish she'd let me cut her nails. They get so long and brittle, then they break off. Better if she'd just let me keep them short, but she won't let me touch them."

"Gert, this pie is delicious. I think maybe we'll go upstairs and see Grandma for a while," Mother said, getting up from the table. She looked at us, motioning to follow her.

"Oh, here. Take this tray up to her, and see if she'll eat something," Aunt Gert said.

Grandpa Wetter had a two-story house with the kitchen, parlor, dining room, and his bedroom on the first floor, and three bedrooms upstairs. There were no bathrooms, just a little house with two seats out back in the apple orchard.

We climbed the tall, narrow stairs to the second floor. Mother went first, then me, then Lenny. At the top of the steps and down the hallway was a room with a locked door. This was Grandma's room.

For as long as I could remember, Grandma had always been locked in this room, for her own safety. Decades later, there would be treatments for Grandma's condition, and more would be understood, but at the time Grandpa was doing the best he could, the only thing he could do.

Balancing the tray in one hand, Mother reached up above the door for the skeleton key, slid it into its hole, and slowly turned it until the door was unlocked. She hung the key back on its nail and opened the door. The sun was shining in through Grandma's window as we walked into her room. Grandma stood up and greeted us in German. Mother responded back in the same language, which, to me, had such a harsh sound. I always thought they sounded angry with each other when they spoke German, though I knew they weren't.

"How are you today, Mother?" Mother asked in German.

"Oh, nothing to complain about," was always Grandma's reply, according to Mother.

Mother set the tray down on a little table next to Grandma's bed. Grandma's meals were served to her on a tin pie plate, her drinks in a tin cup. Glass would have been too dangerous for her to handle in her feeble state.

Grandma walked over to her dresser and picked up her hairbrush, bringing it back to me. She took my hand and laid the brush handle in my palm, then shuffled to the side of the bed and sat down. She waited with her back turned to me. I held the brush firmly as I gingerly stroked Grandma's long, beautiful, silvery hair. She seemed to like this very much, leaning her head back and closing her eyes.

Mother continued to talk to Grandma in German, and Lenny and I didn't understand a word of it. If I had to guess, I'd say she was telling about Old Ted dying, and about the Sunday sermon. Grandma just listened, enjoying our company and the sound of our voices. She was a very gentle lady; clean, well cared for, and always dressed in a nightgown.

After awhile, Mother must have said something about leaving, because Grandma began to cry. This was normal, though. She had a hard time controlling her emotions. We slowly backed out the door, Mother reaching for the key and locking Grandma's room once again. On our way down the stairs, we could hear Grandma chattering to herself in German once again. She was okay now.

"Did she eat anything, Maggie?"

"No, she just wanted her hair brushed. Maybe she'll eat later."

"I hope so. She's shriveling up like a prune. Soon there won't be anything left, you know?"

Mother nodded her head, and raised her eyebrows as if to say, "What can we do?"

Just then, Daddy and Grandpa Wetter came in the door. Grandpa walked in first, offering a short hello without a smile to everyone.

They had been out in Grandpa's shop, where he and Uncle Pat could fix any piece of machinery, make handles for axes, you name it. There was a forge in the shop, used for heating metal. Sometimes Grandpa would let me turn the blower handle to intensify the fire's heat so he could work a piece of iron. Grandpa also had a two-horse buggy in the shop.

He and Uncle Pat farmed with horses just like we did, but Grandpa was on the leading edge when it came to modern conveniences. His was the only farm in the area that had electricity. Grandpa generated his own energy by running a gasoline engine that charged a bank of batteries, and the lights ran off these batteries. They had one bare bulb in each room in the house, and one in the barn and the shop, too. The gasoline engine wasn't running all the time, only when they wanted to use the lights at night.

I was always a little afraid of Grandpa Wetter. He seemed rather stern at times, but he did have a difficult life, with so many

Margie was about thirty-nine years old when Leona took this photo of her mother's family. Grandma had already passed on when her husband and children gathered at Grandpa Wetter's farm in the winter of 1937-38. From left to right: Al Wetter, Pat Wetter (back), Grandpa Wetter, Aunt Gert (back), Margie, and Len Wetter (back).

responsibilities on top of caring for Grandma. I had seen Grandpa's wedding picture. He had been a handsome man at one time, with dark hair, a sharp widow's peak, penetrating, steel-blue eyes, and a long mustache. In the picture, he was wearing a nice suit and standing next to Grandma, not much taller than she.

Even though I had a healthy fear of Grandpa, I never hesitated to show him my report cards, hoping for a small reward for my hard work.

"Hello, Leona," Grandpa said first, with his heavy German accent.

"Hi, Grandpa. I have something to show you," I said, as I handed him my report card.

"Mmmm, hmmm. Looks good. Good grades, mmmm, hmmm," was all he said, reaching into his overalls pocket and pulling out a shiny, new quarter. He took my hand and placed the quarter in my palm, pressing it in firmly, bending my fingers over it. He held my hand for just a moment, then gave me a nod of his head, and I knew he was pleased.

Then Grandpa began to talk with Mother in German, so Lenny and I just sat at the table, listening. Daddy didn't know the language, either, so he turned to me and asked to see my report card.

"How come I haven't seen this yet?"

"I've just been saving it to show to Grandpa first. You want to see it now?"

"Well, of course."

I handed the card to Daddy and he studied it for a minute or two.

"You got good grades, Leona. I'm proud of you," Daddy said, as he patted the back of my hand. When Daddy smiled at me, my whole world seemed a better place, sunnier. I knew I was loved and cherished.

Just then, we heard someone drive up. Lenny and I ran to the door.

"It's Uncle Al! And he's brought Arnold and Lyle with him!" I said excitedly, running outside to greet them.

"Hi, Arnold! Hey, Lyle!"

My cousins greeted us heartily as they climbed out of the car.

"I didn't know you guys were coming today," I said.

Leona's father, Ralph Ahnen, poses with his brothers-in-law, in what may be the last photo to show them all together. From left to right are: Al Wetter (Margie Ahnen's oldest brother); Wilson Shimpach (Aunt Gert's husband and Frank and Emma's son); Pat Wetter (Margie's youngest brother); Ralph; Len Wetter (Margie's middle brother); and Donald Wetter (Len's son and Ralph's nephew). This photo was taken at Grandma and Grandpa Wetter's farm, and in the background you can see the "Pine Knob" mentioned in this chapter. It was shot on Easter Sunday, about a month before the events described in this book's last chapter occurred.

"Yeah, we're going huntin'. You comin' along, too?"

"Well, yeah, I guess. But Daddy didn't say anything about hunting before we left for church this morning, so I don't have my gun."

"That's okay. You can use mine once or twice, if you want," Lyle offered.

He knew I was good with a gun, that I knew how to handle them carefully. Daddy and Grandpa came out of the house.

"Hi, Al," Daddy said cheerfully. "Ready to go huntin', guys?" he said to Arnold and Lyle.

"You bet!" they both cheered at once.

Daddy walked over to our car and brought back his rifle. Grandpa went back into the house to get his.

"Oh, this'll be fun, guys!" I said, so happy to be going with the men. There was nothing I liked better than going hunting with Daddy, except maybe riding my horse.

Al climbed into the driver's seat, Grandpa climbed in next to him. Daddy, Arnold and Lyle climbed in the back. The car was full, and I wasn't sure where I was going to sit.

That's when Daddy dropped the bomb.

"Oh, Leona, not this time darlin'. There's not enough room in the car. Don't worry though. You and I'll go some other time."

Crushed, I slowly stepped away from the car. My heart started pumping faster and my face felt hot. I held up my hands, saying, "Oh, that's okay. You guys go ahead and I'll... I'll just hang out here."

But inside, my heart was broken, aching. This wasn't fair! Why did my cousins get to go, and I didn't? As the car drove off down the driveway, I just stood there and watched, not wanting to go back in the house with the women because I wasn't a woman yet, and didn't want to be one anyway.

Secretly, I kind of suspected that Daddy had always wanted a boy when I was born, but he always seemed glad to have me tag along when he was hunting, trapping, fishing or working in the fields. And tag along I did. I followed him everywhere. I thought this qualified me to go anywhere he went, but I guess I was wrong.

What was I going to do now? I couldn't go back into the house. Mother and Aunt Gert would ask me why I hadn't gone hunting. What would I tell them? That I couldn't go because there wasn't enough room in the car?

This was terribly embarrassing and distressing for me. I just knew the real reason they left me behind was because I was a girl. Life just wasn't fair. Why wasn't I born a boy? And then Mother would tell me that I should be doing more girl things anyway, that I shouldn't be tagging along all the time with the men and boys. I didn't want to hear that, or see the knowing looks on her face. So I headed for the pine knob, off to the woods, where I was free to be myself and sulk for awhile.

As I walked past the house, I looked up to Grandma's room. She was pacing back and forth in front of her window. I could see her mouth moving as she talked to herself, and watched her for a few seconds. Just as I was about to turn away she stopped, put both hands on the window and looked directly at me.

Suddenly, I felt great compassion for Grandma, and guilt about my selfishness.

At least I wasn't locked up. I was free to roam, free to dream, to think. I was free to go to church, even though I didn't appreciate it. I was free, most of the time, to tag along with Daddy. If I really thought about it, I had a pretty good life, unlike poor Grandma; stuck in that room, trapped in her own mind. Maybe my life wasn't so bad after all...

"Wanna hear a ghost story?" Norman asked. "A long time ago, in this very spot, lies a tale of suspense, mystery, and... ghosts."

Lenny scooted closer to me 'til our knees were touching. I moved over an inch, just enough that our knees wouldn't touch. Norman continued.

"By the way, this is a true story, you know."

TALES IN THE WOODS

"Ante-I-over!" I yelled, throwing the ball over Uncle Anton's house and then holding my breath, waiting for someone to come running around the house and chase after me. I was a fast runner, and so far no one had caught me all night, not even my cousin Norman. Suddenly, Norman came sneaking out of the bushes behind me and tagged me with the ball, before I'd barely had a chance to run.

"Norman! How did you get there? Were you on the other side when I threw the ball, or have you been over here the whole time?"

"Now, how could I get the ball if I wasn't on the other side of the house to start with?" Norman answered.

"Well, I don't know. Then how did you get over here so fast?"

Norman was rolling on the ground, laughing.

"I got Leona! I got Leona!"

"Norman, stop it! I'm not playing if you're going to be cheating."

Norman got up from the ground and sobered up a bit. By then, Lenny, Kathleen and Gen, Kathleen and Norman's little sister, were gathered around the two of us.

"Don't worry Leona. You're just sore 'cuz nobody can catch you all night, and all of a sudden I outfoxed you!"

"Outfoxing's one thing, but cheatin's another!"

"For your information, I waited 'til the ball came over, and then I ran into the woods and came out behind you in the bushes." I stole a glance at the woods, the house, the bushes, and realized it was entirely possible."

"All right, I believe you this time. But I say we do it over," I said, trying to recover my dignity.

"Okay," Norman agreed, reluctantly. "Come on, Lenny and Gen. Let's go back to our side and we'll beat them again," Norman encouraged his troops.

Kathleen and I huddled together.

"This time someone's got to watch our backs, 'cause Norman's getting sneaky," I said.

"I'll do it, 'cuz you can run faster," Kathleen volunteered.

We broke our huddle. I picked up the ball and whipped it over the roof again, yelling, "Ante-I-over!"

Kathleen kept a close eye on the spooky woods while I watched both sides of the house. Just then, Lenny and Gen came walking nonchalantly around the corners on both sides of the house. It startled me for a minute, but neither of them had the ball. I cautiously ran around to the other side of the house, stopping by the front door.

The kerosene lights from the house glowed out into the dark, fall night. It must have been well past nine and we were still going strong, running, playing and fighting 'til our cheeks were flushed and we had that clean, outdoors smell of fresh air and exertion.

It was always exciting to go over to Norman's house, and we did visit them often. Daddy would tell us all to hop in the wagon, we're going to Uncle Anton's!

We'd never heard of a babysitter. We just simply went where our folks did, and went home when they did. If we got too tired, we would find a place to curl up and go to sleep until Daddy and Mother were ready to leave.

I could see Daddy and Uncle Anton sitting at the kitchen table, having a heated discussion. Mother and Aunt Blanche were talking, too, probably about the usual things: children, gardens, neighbors, and the last episode of "Ma Perkins."

All of a sudden, I heard something moving in the rosebushes next to the house. I moved in a little closer to see what animal would be hiding in there.

"Yaaahhhhh! Got ya again!" Norman yelled as he tore out from the bushes, scratching his arms and face as he lunged for me, and knocking me down hard.

"Darn, Norman!" I yelled back.

Lenny happened to see the whole thing. She sucked in her

breath and said, "Leona, you're not supposed to say that. I'm telling Ma."

"Well, Daddy says it sometimes and he doesn't care, so don't you be telling her!"

Lenny rolled her eyes and kicked at a stone.

"Come on, guys, let's go inside for a while. I'm thirsty," Kathleen said.

We all followed Kathleen, filing into the house one at a time, finding a spot on the kitchen floor where we could sit and listen to the adults.

"Want something to drink?" Kathleen asked. We all nodded our heads. Kathleen left to get the milk from the pump house near the windmill. When she returned, she poured each of us a glass, and we listened to the adults talking by the light of a kerosene lamp.

"Ralph, I'm tellin' ya, there's a war comin' on," Uncle Anton said seriously.

Daddy just shook his head and said, "Yeah, well, maybe. We'll see what happens with the Germans over in Europe first. It'll probably all blow over in no time."

Uncle Anton wasn't so sure. "Those Germans, from what I've read, have quite an army, big tanks, and Hitler…"

"Daddy," I interrupted, "There won't be a war, will there? You wouldn't have to go fight, would you?"

"See, Anton?" Daddy said disgustedly. "You're scaring the girls, and there's not much we can do about it, anyway."

Uncle Anton shrugged his shoulders and said, "You heard anything about guns being stored in the basement of the Catholic church?"

"What? Now where did you hear this?" Mother interrupted. "Never mind, don't answer that. I don't want to know."

Daddy changed the subject.

"You coming over tomorrow so we can do some more work at the quarry?" he asked Uncle Anton.

"Yeah, I suppose so."

Daddy and Uncle Anton were always trying to find a way to earn more money than just farming, especially Uncle Anton. He had a Fordson tractor, and between the two of them, they somehow managed to buy a rock crusher. They used the tractor

Taken sometime during the early years of World War II, a ten- or twelve-year-old Lenny poses in the Ahnen's buggy with their Fox Terrier, Penny. The buggy sits on the road above the family farmhouse, and the trees in the background are on Hickory Hill, where Leona and Norman did their gopher hunting.

for power to run the crusher. There certainly was no shortage of lime on most of our fields, and they knew they could sell to other farmers if they could produce. There was a huge outcropping of limestone rock on our road, and here they started digging.

It was backbreaking work, hot and noisy. First, they cleared away the topsoil, which was done with a team of horses and a scoop. Then they set dynamite and blasted some of the rock loose. These rocks were loaded by hand into a wheelbarrow, then pushed up a ramp to the top of the crusher. From there, the rocks were dumped on a platform, and one man fed them into a hole where they fell onto the whirling hammers, which pulverized the rocks into fine lime. The lime went up a conveyor, where it was either dumped into the back of a truck, or in a pile on the ground.

Mother didn't think Daddy and Uncle Anton were getting very rich with their venture into the commercial realm, but they sure dug a mighty big hole in the cow pasture!

After listening to talk of limestone and war, we decided to go back outside.

"Hey, guys, I got an idea," Norman said excitedly. "Let's walk down to the old home place."

"Yeah, let's go, but we'll need a lantern," I said. "There's no moon tonight."

Norman ran to the barn and came back carrying a brightly lit kerosene lantern.

"Let's go, girls," Norman said, walking away from the barn.

I followed first, then Lenny and Gen, and Kathleen brought up the rear. The path led us away, further and further from the barn, to a steep hill. Then we started down the other side, trying to keep our momentum under control. We dug our heels into the ground, holding onto branches wherever we could, to keep ourselves from sliding. We knew if we let ourselves go and just tried to run for the bottom, we'd end up tumbling head-over-heels and would most likely get hurt.

It was hard to see, and Norman's lantern didn't do much good when he was so far ahead of the rest of us. Finally, we felt the earth beneath us level out somewhat and we continued down the valley, twisting and turning until we got to the old home place.

The only remains left standing from Grandma and Grandpa Ahnen's original homestead were a few logs from the old barn, and the stone fireplace from the log house. This was where Daddy and Uncle Anton had been born, along with Margaret, Kate, Leona and Ernest. Norman set the lantern on a fallen rock at the base of the stone fireplace.

"You guys wanna hear a ghost story?" Norman asked.

"Sure! Let's all sit around in a circle on the ground," I said excitedly.

Norman sat right in front of the fireplace. To his right sat Gen, then Kathleen, Lenny and me. He began, "A long time ago, on this very spot, in the log cabin where Uncle Ralph and Dad were born, lies a tale of suspense, mystery, and... ghosts."

Lenny scooted closer to me, 'til our knees were touching. I moved over an inch, just enough that our knees wouldn't touch. Norman continued.

"By the way, this is a true story, you know."

"No way, Norman. I've never heard Daddy tell us some true ghost story about this place before," I said.

"Well, maybe he didn't want to frighten you, but I swear it's true. You can even ask my Dad if you want."

I looked over at Kathleen for some telltale sign of conspiracy, but she just nodded her head.

"It's true. Dad's told this story to us many times, and he claims it's absolutely true."

"Well, okay." Lenny scooted closer to me again. Again, I moved away. All eyes were on Norman.

"The story begins like this: One day Grandma and Grandpa Ahnen had to go into town, so they took the horse and wagon, leaving the older children in charge of the younger ones back here on the farm. Everything went fine that day, until evening came on. It was a dark, moonless night—just like tonight, probably the day before Halloween—and Grandma and Grandpa weren't home yet.

"The older kids were getting a bit worried. They wondered if perhaps their folks intended to stay in town all night, but no one had said anything about being gone overnight. Then they thought something horrible must have happened to them on their way home. They just knew it. The younger ones didn't realize the seriousness of the situation and were misbehaving, being too noisy and causing a ruckus. Most likely your Daddy and Uncle Ernie," Norman said to Lenny and me, with a twinkle in his eye.

"Oh, right," I said.

"Well, anyway, some of the kids were getting into trouble. No parents around, it was dark outside... Anything could happen."

This time, Lenny scooted right into my lap. I decided to let her stay.

"Suddenly, a ghostly apparition appeared at the top of the steps going up to the loft," Norman said, in a spooky voice.

"What's a apparition?" Lenny whispered to me.

"A ghost," I whispered back.

"Yes, there at the top of the steps," Norman said, pointing behind himself into the sky, where the steps might have been at one time. "Right up there stood a ghostly lady, all white and glowing, just hovering at the top of the steps in the loft. Suddenly she spoke, and she said, 'Hush, my baby wants to sleep,' and then disappeared."

Lenny burrowed her head into my chest. Gen's eyes were circular moons, and Kathleen looked spooked, as well. I know I was spooked, especially sitting out in the woods, in the very spot where it had happened! I kept thinking that, at this very spot where we sat, at one time, a ghost had actually appeared. That is, if Norman wasn't pulling my leg.

"That's the end of the story. Creepy, huh? But absolutely true, I swear," Norman said in all seriousness. Suddenly, an owl hooted and we all jumped in our skin.

"Leona, I'm scared. I wanna go home," Lenny pleaded.

"Oh, don't be a baby," I said. "It's just a story."

"But Norman says it's true," Lenny said.

"Lenny, whatever happened here occurred a long time ago, when your Daddy was just as small as you. I'm sure the ghost is long gone by now," Norman said.

Not wanting to be outdone, I decided to put in my two cents. "You guys have heard about Black Jack Bachim, haven't you?"

"Yeah, of course! Who hasn't?" Norman said.

"Is he the black guy that works for the railroad? Always wears overalls, a felt hat and a red handkerchief around his neck?" Kathleen asked.

"Doesn't he repair the tracks or something?" Norman added.

"Yeah, that's the guy, and I think he does repair the tracks. I think he rides on one of those railcars that has a pump handle in the middle, and he pumps the handle up and down to make the car go."

"What about him?" Gen asked.

"Well, are you ready for this? I don't know if I should say."

"Oh, come on. You gotta tell us, now!"

"Okay. Did you know that we're related to him, he's actually kin to us?"

"What? Are you kidding? Who told you this?" Norman demanded.

"I'm serious. Daddy told me. He's related to the Ahnens, by marriage."

"Wow! How come I never knew this?" Norman asked.

"I don't know. Maybe most people don't know it, but Daddy did and he's the one that told me. Said he's right proud of being related to Black Jack because he's a good, hardworking man."

"Leona, I don't believe you. I'm telling Mother you told a fib," Lenny said.

"It's the honest truth, you can even ask Daddy!"

Lenny crossed her arms over her chest and sat with a disgusted look on her face. I could tell she didn't believe me.

"Ah, come on Lenny. Your sister wouldn't lie. And besides, it's kind of neat!" Kathleen said. "Kind of sets us apart from most folks around here. Makes us special, you know?"

"Well, Leona, that's hard to believe, but I know you wouldn't tell a lie, so it's just gotta be!" Norman decided.

"You want to hear another story?" I asked.

"Sure!"

"Well, this happened awhile back, when I was about three years old, I guess, and long before Lenny was born. Daddy suspected someone was stealing corn from us. So one day he put a matchstick in the corner of the corncrib door. That way, if someone opened the door he'd know it, 'cuz the matchstick would fall out. Well, one day we came home from somewhere and the matchstick was lying on the ground!"

Lenny was about six or seven years old when she posed for this photo, dressed as a gypsy. It wasn't for a party or any special event. Lenny simply had a vivid imagination, and liked to dress up and pretend. According to Leona, "She would put on everything she could get hold of."

The tarpaper covering the outside of the house was its usual siding, but that ripped away from over the door tells the tale of the previous hard winter that had blown so hard it forced cold air in around the closed door. Ralph had tacked up the tarpaper in an attempt to close up some of the gaps. When the warm weather arrived, it was simply torn away. There were no door or window screens. The wet porch is evidence of an early spring frost melting under the sun's warming rays.

"Really?" Norman asked.

"Well, then Daddy knew he had a real mystery on his hands. He had to find out who was taking our corn. He had his suspicions, but no way to prove them. So a few days later, he told Mother and me, and Mother's cousin, Lucille, to get in the wagon. Said we're going down the road a ways. We all hopped in the wagon and Daddy drove just a short ways down Dry Hollow road. Then he stopped."

"What then?" Kathleen asked excitedly.

"He said he was going to walk back to the house in hopes of catching the thieves red-handed. So he did, and we just waited in the wagon, not knowing what was happening. Mother said she didn't really want to know because maybe the person who was taking the corn needed it more than we did, and she just didn't like confrontations of any kind."

I looked around the group, and all eyes were fastened on me. I continued.

"So, Daddy walked back to the house. And sure enough, he found someone in the corncrib. And someone in the cellar, and someone in the pantry! They were taking corn, all right, and other food, too. Then Daddy had to call the sheriff to come out and settle the dispute. I guess one of them went to jail for a night or two, and the other two got off with a warning."

"Who was it?" Norman wanted to know.

"I don't know. Daddy would never tell me. He did say he felt kind of bad after it happened, because these folks really did need the food. And Mother wasn't too happy about the whole thing, either."

"That's really strange. But your Daddy's a fair man, and I'm sure he did what was right, hard as it may have been," Norman said.

And with that, he stood up, picked up the lantern, and the rest of us followed without saying a word. We were all getting tired.

Taking the path back home through the twisting, turning valley, away from the old home place, we walked in silence. Climbing back up the steep hill was more difficult than coming down, but somehow we all made it without slipping and falling.

We found our way back to the barn, and Norman blew out the lantern and returned it to its place on the wall. Then we all walked to the house, but instead of going in, we just plopped down under an old oak tree. Eventually, we all fell asleep right there, with thoughts of ghosts and war, thieves and Black Jack Bachim dancing through our minds.

Ralph Ahnen, Leona's beloved father, was thirty-eight years old in this photo. Like most farmers of his region, he was responsible for keeping a ready supply of fuel on hand for the wood stoves with which their wives cooked and heated the house. Here, Cute and Babe wait patiently for Ralph to load tree trunk segments where they were cut onto a sled, which the horses will pull to the yard in front of the house. There, Ralph will cut them into stove lengths and split them for Leona to carry in to fill the kitchen wood box.

One wintry day, Daddy walked in the house, stomping the snow off his boots by the door.

"Say, did I tell you guys I got a skunk and a coon this morning in my traps?" he asked excitedly.

Mother clapped her hands and said, "That's good."

"How much do you think you'll get for the hides?" I asked.

"Ohhh, maybe six bucks or so!" Daddy said.

This was a bonus day for our family. Daddy had a trapline, and during the Depression, coon, fox and skunk hides paid for groceries in the winter.

Winter Happenings

"Tim-ber-r-r-r-r!" Daddy yelled, giving the oak tree one final shove.

The tree, which was about to become firewood, finally gave way and fell to the soft earth with a crash. A foot of snow on the ground certainly helped soften its fall. Daddy began to chop the limbs off with his ax, deftly swinging it with sure, swift chops. Then he cut the tree into lengths he could handle, and I hurried over to help.

"Heave-ho, Leona!" Daddy said, as we lifted the five-foot log onto the horse-drawn sled.

We went back for another and another, until we had a whole load. I scrambled up to the top of the log pile, while Daddy drove the team of horses all the way back to the farm. Once in the farmyard, we unloaded the logs next to Uncle Anton's Fordson tractor, which had a saw mounted on the front. This was used to saw the wood into stove lengths.

After Daddy had cut several of these stove-length pieces, I began to take them, one by one, and set them upright on the ground so he could come around and split each with his ax. One piece after another, with aching arms, cold fingers and runny nose, I worked hard trying to keep up with the tractor. It wasn't long before we had a regular forest of stumps waiting to be split. Finally, Daddy took a break from sawing and grabbed his ax.

Heaving the ax way up over his head, Daddy let the weight of the ax head do the work, as he aimed for the middle of each piece.

Whump! the ax sounded into the cold, winter air. *Whump!*

Daddy moved to the right a few feet. *Whump!* He moved a few feet more. *Whump!*

Like he was on an assembly line, Daddy went from one piece to the next, splitting them in two with one whack of the ax. It didn't take long before he was way ahead of me again. I hurried to pick up the pieces as they fell and stacked them in nice, neat rows, all the time trying to stay well out of Daddy's way and away from the swinging ax.

Soon, all the pieces were split and Daddy went back to sawing again, while I finished picking up the split wood and stacking it. After a couple hours of this, the light began to fade. It was only four-thirty or so in the afternoon.

"Well, let's call her quits for today," Daddy said, piling a stack of wood in my arms to carry up to the house.

I walked through the snow and up to the kitchen door.

"Open!" I yelled politely.

Mother opened the door to let me in, and I staggered under my heavy load over to the woodbox. I let the wood fall into it from my arms, with a loud crash.

"Leona! Please don't do that. You startle me every time," Mother pleaded, a stern look on her face.

My arms ached. I rubbed my sore biceps and at the same time tried to warm up my hands. Mother and Lenny were at the table, setting down plates and silverware for supper. Mother had her back to me. I turned around and reached up with both hands, laying them against Mother's soft, warm neck.

"Aieeee!" Mother yelled, quickly brushing my hands away.

I couldn't help but laugh. This trick worked every time, always getting a reaction out of whomever I decided to torture with my freezing hands.

"Leona, go on now and get some more wood!" Mother commanded.

I slowly dragged my tired body back outside for another load. Daddy eagerly filled up my arms again, and I trudged once more back to the house.

"Open!" I yelled. Nothing happened. "Ooopennn!" I yelled louder.

Mother appeared, looking through the door's window, smiling.

"Mother, open the door!"

She cupped her hand behind her ear and mouthed something from behind the door, looking at me with puzzlement.

"OPEN THE DOOR!" I yelled, even louder.

Mother pretended to suddenly understand, as she ever-so-slowly opened the kitchen door. She was smiling a wicked smile as I walked past her. I didn't appreciate her humor at all. I walked over to the woodbox, lifted my arms up as high as I could stand to, and let the wood fall with a bang. Mother jumped.

Winters could be harsh on the farm. Some winters, we were literally snowed in for months. Even the main road going past the school and out the ridge toward town was impassable. The road itself was, at times, below the surrounding fields because, with fencerows full of brush and trees acting as a snow fence, there could be huge drifts along the road.

Sometimes, if there was a heavy crust on the snow, we could walk on that to get to school, or else we walked at the edge of the field where the wind blew the snow away.

One winter, all the men in our neighborhood shoveled ahead of the snowplow until the ridge road was opened. Our mailman, Henry Everson, always made it through, no matter what. He would have someone drop him off where the road was drifted over and walk five or six miles from there to deliver the mail. Then he'd be picked up again, closer to town.

Another winter, we were not only snowed in, but quarantined as well.

I was the first in my family to get scarlet fever, having caught it at school. A large, yellow plaque was placed on our front door, warning people not to enter this house. The sign stayed in place for about a month. Then Lenny—a tiny baby at the time—got the fever, too. So another month went by when no one could come into our house. Then Mother and Daddy in turn also contracted the fever, and we were quarantined for yet another two months.

Daddy became very sick, talking out of his head, which really scared Mother and I. But eventually, the sign came down and the house was fumigated with an awful-smelling disinfectant. Then we were fit for society again.

I don't remember ever believing in Santa, but it was so much fun to have a little sister who did, and I never let on. Grandpa Wetter always brought us a Christmas tree cut from the pine knob. It smelled so wonderful, and what fun we had trimming the tree! We had some beautiful old ornaments, and we strung popcorn and bittersweet to hang on the tree. We also had some folding, red paper bells that we hung from the ceiling. We put real candles on the tree and were allowed to light them at night.

Christmas Eve was a special night on our farm. We always gave the cows and horses extra silage and grain, and it seemed that they, too, knew it was a special night.

After milking the cows, Daddy and I would walk back to the house, our boots crunching on the new snow, sticking our tongues out, trying to catch the large snowflakes as they fell from the sky. Supper would be ready and we'd have our own little feast. But it was after supper that the real party began: It was time for presents!

Lenny and Mother and I waited for Santa. Lenny never seemed to notice that Daddy was missing. He would bang on the outside of the house and Lenny would squeal with delight, "It's Santa! He's here, he's here!"

Then Daddy would come in and we'd open our presents.

Money was scarce during the Depression, but despite this fact, one year I received a Brownie® camera, another year some oil paints, another year a pair of wood skis. I knew my parents had to buy our presents and couldn't believe that they would spend all that money on us!

One year, the Penney's store in Boscobel advertised that they had Santa's workshop in the basement. I didn't believe in Santa and don't know what I expected to see, but I was pretty disappointed and mad when I found out it was just a store to sell toys, dolls, and other stuff. It wasn't a workshop at all!

One wintry day, Daddy walked in the house, stomping the snow off his boots by the door.

"Say, did I tell you guys I got a skunk and a coon this morning in my traps?" he asked excitedly.

Mother clapped her hands and said, "That's good."

"How much do you think you'll get for the hides?" I asked.

"Ohhh, maybe six bucks or so!" Daddy said.

Taken around 1937, this photo shows Leona at about age twelve, and Lenny would have been around seven. Lee holds a pair of skis, a Christmas gift from her parents, while Lenny holds barehanded onto a steel-runnered sled, a must-have for any Wisconsin youngster to make the most of the long winters.

The wooden structure in the background is an abandonded rabbit hutch Leona once used to keep her pet black rabbit in. She had brought the rabbit home from Uncle Al's house one day when she and Norman had gone over to play with Lee's cousins. Her aunt had given the rabbit to her in a gunnysack, and she had carried it that way for five or six miles, all the way home.

This was a bonus day for our family. Daddy had a trapline, and during the Depression, coon, fox and skunk hides paid for groceries in the winter. Sometimes, I'd get to go with Daddy on the trapline, and this was always a treat for me.

Daddy also loved coon hunting and had a couple of Walker Hounds, Jack and Old Drum. These two dogs lived for coon hunting. But they also made good "horses" when I needed to "plow the driveway". I'd make a harness out of binder twine, hitch the dogs up to Mother's garden hoe, and we'd "plow" the sand in our driveway, making believe we were doing the real thing.

During the early spring and fall evenings, when the dogs weren't plowing the driveway for me, Daddy and some neighbors loved to let their hounds run, just to hear the music of their baying. I never went along on the coon hunting trips, but I do remember listening late at night to Daddy telling Mother all about the chase; where the fox went, how it doubled back, how the dogs sounded. She listened to all he told her, not really understanding a lot of it, just listening to the story. But Mother did know the value of these animal skins. Animal hides meant money for food.

After a long day of walking the trapline, or "making wood" with Daddy, I'd be bushed. We'd all gather in the dining room to listen to our favorite radio program, "The Lone Ranger." Well, actually only Daddy and I tuned in. Lenny was frightened by all the gunfire and mayhem and the music, so she and Mother would stay in the kitchen with the door shut, while Daddy and I rode off in our imaginations on the back of the most beautiful white stallion, Silver.

The Lone Ranger was always on the side of good, with the help of his Indian sidekick, Tonto, and they won out over evil. Daddy and I would sit, glued to the radio station, every Wednesday night at seven o'clock, as the music began to echo through our battery-operated Coronado radio, which we'd bought from the Gamble's Store. This radio had a wet battery that needed to be taken to town for recharging when it ran down, and several other dry cell batteries that could be replaced.

We also listened to the WLS Barn Dance from Chicago on Saturday nights at seven. We were entertained by such talents as Lulu Belle, Scotty, Georgie Goebel, the Williams Brothers, and

Red Foley, who sang "Old Shep," a sad story about his old dog. Lenny always cried when she heard that song. Little Jimmy Dickens sang, "Take A Cold 'Tater And Wait," a song about a little kid who was hungry and had to wait for the grown-ups to eat first, so his mother told him to take a cold potato and sit down and wait!

We heard the same entertainers every show and got to know what to expect from each one. There would be square dance music and the whole cast would participate. Once a year, we could send to WLS to get an album of the Barn Dance, which cost about a dollar. It showed pictures of the stars and some of the songs and music. There was a program on WMAQ from Chicago called "The Grand Ole Opry," but we never listened to that. We were loyal to our WLS show and thought of the people there as friends and family.

Mother liked to listen to "Ma Perkins," which had to be the first soap opera. They really did sell soap, from their sponsor, Oxydol. I couldn't have cared less about Ma Perkins, but as soon as we got home from school at four-thirty, we tuned in to Jack Armstrong, the "All American Boy," cowboy Tom Mix, and a couple other programs for a whole hour before doing chores. This was our way of winding down after a long day at school.

These stories ran fifteen minutes, and were sponsored by Wheaties®, Ralston®, and similar products aimed at children. Lenny and I saved box tops to send away for telescopes and walking meters. Each offer also required a quarter, which wasn't always easy to come by.

The radio surely filled a big hole in our lives, particularly in the wintertime, when we were so hungry for music and the wholesome songs and nonsense skits. The images we pictured were more wonderful than any color TV could ever be, during those cold winter nights in our small farmhouse.

If we weren't listening to the radio, we were reading books. We all loved to read. Mother and I belonged to the Sears "People's Book Club," and once a month, we got a new selection. They were all good reading, novels and adventure stories. Mother always read the ending first and if she liked the ending, she knew she'd like the rest of the book. Daddy loved westerns, which were pulp magazines printed on coarse paper, and cost about ten to fifteen cents each.

Leona took this photo sometime in the winter of 1937-38, trying to capture the wondrous effects of an ice storm that had ravaged the Boscobel area. The ice was hanging from the trees, and everything was covered with it, beckoning to the little girl with her new camera. The boards around the yard's perimeter acted as a retaining wall to keep the yard higher than the surrounding area. This was important to a fairly flat area where rain could quickly turn packed dirt to mud and manure piles to rivers of odorous ooze. The fence was a permanent fixture made of woven wire, strung around the yard to keep out the cattle, horses, and other livestock, which would quickly denude it of any remaining grass. The horseshoe enmeshed in the lower right corner most likely was placed there by someone who got tired of stepping on the stray hardware and finally picked it up and hung it on the fence.

There were usually three or four long stories and a couple of short ones in each magazine. Daddy had stacks of these, most of them given to him by friends.

In the winter evenings, we would read by the light of a kerosene lamp. Daddy would sit in his chair, and before he became a Christian, he'd pull out his pipe, tamp a little tobacco in it and light it, then settle in to read the long stories written in the magazines.

I couldn't understand the plot in those long stories, so I read the short ones, instead. They were all about cowboys and Indians. The white men were called *hombres*, and the bad men were called "greasers." They chased each other through the desert, and dry-gulched each other from the arroyos. There were some very bad men.

One story both horrified and fascinated me. The hero of the story was captured by some bad greasers, then sewn into a wet rawhide blanket, and left in the hot sun. The rawhide would shrink, squeezing the hero to death, and killer ants were closing in on him. I'm sure he was rescued in time, but I shuddered every time I thought of that story. It was worse than some of the Mother Goose stories, or even Hansel and Gretel.

A couple of my favorite books were two of Gene Stratton Porter's, which I read over and over. One was "Girl of the Limberlost," and the other was "Freckles." They both had much in them about flowers, gardens, birds, forests, and other natural things. They influenced me to build a wildflower garden back in Gunderson's Woods one summer.

I cleared a spot under a big oak tree, hauled water from a spring back in the woods, and transplanted violets, ferns, wild geraniums, honeysuckle, and, of course, ladyslipper. I spent many hours out there all by myself, and learned an important lesson: If you have a secret, better to keep it!

I took Lenny and Mother to see my garden one day. I'm sure Mother appreciated it, and Lenny did nothing to harm it, but after that, it was not as important to me anymore. Since it was no longer my own special secret, I lost interest, but I never lost my fascination with flowers. During the long, cold winter months, reading about flowers and birds helped to keep my spirits up.

After reading for an hour, Daddy and I would become hungry. Our favorite snack was "broke-up-bread." We put pieces of homemade bread in a bowl, put some sugar and milk on it and ate it up. After that, we'd read one last book, my all-time favorite, "Little Black Sambo." Then Daddy would say, "Time to hit the hay, girls."

Off to bed we'd go, to literally fall, exhausted, into our straw beds.

Daddy paused, his eyes wide open, not even cracking a smile. He pretended to cautiously climb up the hill, using his arms and hands.

Lenny crouched down in Daddy's lap, trying to stay out of harm's way. The suspense was exciting.

The Trouble With Lollygagging

Lenny was sitting on Daddy's lap with her round, pudgy face lying against his chest, the buttons of his overalls making an imprint on her cheek. Reaching into Daddy's shirt pocket, Lenny pulled out his Prince Albert tobacco and held it under her nose. My little sister and I liked the sweet, musty smell of Daddy's tobacco, unlike Mother, who thought it was a nasty habit. Daddy always liked to tell the joke about Prince Albert tobacco.

"You ever hear about the time I walked into the store and asked the clerk if he had Prince Albert in a can? And the store clerk said, 'Sure do!' So I said, "Well, what's he doin' in there? Better let him out!'"

Lenny and I always laughed at Daddy's jokes. He had such a good way with making us laugh, and also making us feel that we were important and could really be somebody someday. We believed him.

Daddy was resting in his favorite chair, just staring out the window at the snow-covered hills of Boscobel country engulfed in semi-darkness, but with the full moon shining. The day was winding down now that the milking was done, the cows, horses, chickens and hogs were fed, and Mother was doing dishes in the kitchen. A lingering aroma of fried chicken hung in the air.

I watched as Daddy absently reached up to take the can of tobacco out of Lenny's hand and put it back in his pocket without saying a word. Lenny had an earache, so Daddy gently blew smoke in her ear to make it better, and told her all kinds of made-up, silly cow stories. It seemed to be working. Lenny was relaxed and resting quietly in Daddy's arms.

"Daddy, tell us the story about Aunt Kate and the wild Indians," I said, interrupting the silence. I loved to hear the stories Daddy told of his childhood days.

Leaning back in his chair and running his large, callused fingers through his hair, Daddy began to remember.

"Well, the story of Aunt Kate and the wild Indians happened after we moved into the new buildings up on the hill from the old homestead in the valley."

I could picture in my mind the remains of Daddy's boyhood cabin home in the valley, the place his Daddy had built when they first came to this land. The only thing left was the old fireplace. The logs and everything else had disappeared into the earth, gone back to the soil they once came from. It was kind of a spooky place to be, knowing that my ancestors once lived on that spot; that, at one time, they ate, slept and worked there.

My grandmother gave birth there many times, and children's feet once trod those grounds where they played, cried and worked alongside the adults. To most folks, the place would have meant nothing, but for Daddy it was almost like sacred ground.

Those days are long since past, and the sounds, smells, and sights of the people who once inhabited that place are being buried along with the rotting logs, except for selected stories that have been passed on to me, and which live on in my mind almost as if I had been there myself.

"And it was up to us kids to bring home the cows from the pasture every afternoon," Daddy continued. "The hills in the pasture were very steep, and the valley below was kind of winding and twisted. This pasture covered at least a hundred acres. Anyway, we took turns bringing the bossies home. One day us boys would do it, and the next would be our sisters' turn, and so on."

Lenny was sitting upright on Daddy's lap now, watching his face and the expressions he made while telling the familiar story.

"One afternoon, it was the girls' turn to bring home the cows, and they were heading down toward the valley, kind of lollygagging."

Daddy paused. I knew what was coming.

"What's lollygagging?" Lenny interrupted.

She did this every single time! Daddy had told her at least a hundred times what lollygagging means, but every time Lenny had to ask and Daddy patiently explained it to her, yet again. This ritual had become part of the story.

"Well, lollygagging is when you're kind of goofing off and not doin' what you're supposed to be doin', see. But then again, maybe lollygagging's what you're supposed to be doin', but you just don't know it."

Lenny looked puzzled, and so was I. That last part was new to us. Apparently, Daddy had turned this idea over in his mind some since we'd last heard the story.

"Like Leona does when Mother tells her to do the dishes?" Lenny volunteered.

"Daddy, I do not!" I protested, trying to defend myself.

"It's okay, girls. Lollygagging's not such a bad thing if you got the time to do it. I did my share of lollygagging when I was a boy. Hey, it's just part of being a kid, right? Unless you're supposed to be working, Leona."

I shot Lenny my "you're going to die" look, but she pretended not to see. Lenny was such a pest. Mother and Daddy thought she was so innocent, but I knew better.

"Anyway, my sisters were taking their sweet ol' time heading down the hill. The cows were down in the valley and the girls were in no hurry to round them up. They were just a-dancing and singing, picking flowers, totally oblivious to us brothers, spying on them from up above, on the ridge."

"How come is it all right for boys to spy but not girls?" I asked, rather indignantly, a hint of challenge in my voice.

"Oh, we were just boys having some fun, but for a young lady to be spying, well... that's not proper, Leona," Daddy said, giving me a wink and causing my resolve to melt.

I let my case rest, for the time being. Somehow, Daddy had a way of taking the wind out of my fire in such a kind and teasing way. I guess he didn't take me too seriously.

"So here we were, lying on our stomachs at the top of the ridge, looking down over the pasture into the valley below. Kate, Margaret and Leona were about halfway down the hill, when I got this bright idea. We backed down a ways on the other side of the ridge to discuss our plan."

Leona took this shot of Lenny at about seven years of age. Like all her family members, Lenny had a tremendous imagination. While Leona and her father listened to Tom Mix on the radio and immersed themselves in pulp westerns, Lenny often indulged herself in playing "dress-up." Here, she's ready for a game of "Cowboys and Indians," which she played with her older sister quite often.

I sat up on my heels, ready for the next turn of events. Lenny had her hand up to her mouth, concealing her wide grin.

"Anton, Ernest and I, we took our shirts off, and rubbed dirt all over our faces and chests. Mixed with a little spit, it kinda looked like war paint. We rubbed dandelions all over our faces, messed up our hair, and then we slowly crawled up to the top of the ridge."

Daddy paused, his eyes wide open, not even cracking a smile. He pretended to cautiously climb up the hill, using his arms and hands. Lenny crouched down in Daddy's lap, trying to stay out of harm's way. The suspense was exciting.

"We get up to the top and carefully peek over the ridge," Daddy whispered. "There they were, still halfway down the hill, making daisy chains or something. While the girls had their backs turned, we jumped over the ridge and came running down the hill, waving our shirts in the air, hollerin' and hootin' like wild Indians!" Daddy said, his words rushing out faster and louder.

"Ooh-boo, boo, boo!" Lenny yelled, her hand rapidly patting her mouth as she tried to sound like a wild Indian. Daddy laughed and continued.

"Well, the girls jumped up and turned around, saw us coming down the hill more than a hundred yards away, yelling our fool heads off, and they just turned tail, making a run for the valley as fast as they could go!" Daddy hollered as he clapped his hands together and slid his right hand forward, as if pointing us in the downhill direction.

"As they were runnin' for their lives, your Aunt Kate was doin' the darnedest thing; darting to the left, then to the right, then left, then right, waving her arms wildly above her head," Daddy explained, using his arms to demonstrate. "We couldn't figure what on earth she was tryin' to do! Had she gone mad? Did we frighten her so bad that she lost her wits? We were stumped," Daddy said with finality and stopped the story abruptly.

"Keep going! Keep going!" Lenny said, so Daddy continued.

"But then, all of a sudden, as she was careening down the hill, a-ziggin' and a-zaggin', she screamed to her sisters, 'Zig-zag so they can't hit you with their arrows!'"

Lenny and I laughed and hollered as Daddy played out the scene.

"The girls were so scared, they didn't even stop to see who it was that was chasing them, so terrified were they of being run down by some savages who would capture them and force them into brutal slavery for the rest of their lives!"

Daddy grabbed Lenny with a big bear hug.

Lenny and I laughed 'til our sides hurt, repeating Aunt Kate's famous line, "Zig-zag so they can't hit you with their arrows!"

Daddy was laughing too, for it seemed that as many times as he'd told that story, he still got a kick out of it.

"Were there any real Indians back then?" I asked, wanting to know more about the "good old days," but Daddy was already standing up to take Lenny off to bed.

"Come on, Leona," he said, "Time to hit the hay."

I never did get an answer to my question.

That night, while I lay in bed, snuggled down under our heavy, wool blanket and listening to Lenny's soft breathing and the wind blowing outside, I pictured myself on our horse, Old Ted. Well, maybe a young, spirited paint instead. I could see myself riding through the tall prairie grasses of the Plains, riding west toward the Rocky Mountains; just me and my horse, a bedroll, a couple cans of pork and beans, ammunition in my saddle bags, and my gun in its scabbard, riding away from home in search of adventures unknown.

And way off in the distance, I could hear a primitive drum beat, echoing in my soul: "THUM-thum-thum-thum, THUM-thum-thum-thum..."

Grandma Ahnen lived in the house where I was born, the "home place." Over the years, the home place had a succession of caretakers. After Grandpa and Grandma Ahnen ran the farm for many years, Daddy, Mother and I lived there until I was two years old, when we moved to the old Benson place a mile and a half away. After we moved, Uncle Anton and his family ran the farm, then Uncle Ernie and his family. Grandma had two rooms all for herself, and did her own cooking.

Grandpa Ahnen had died of cancer a few years before I was born, so I never knew him. Grandma said he had worked in the iron mines in Iron River, Michigan after coming to the United States from Luxembourg. Then he married Grandma, Mary Bachim, and they had settled on a two-hundred-acre parcel of land in Hickory Grove Township. Their six children, including my father, had been born in a log cabin they built in the valley of this land, but they eventually moved the whole family out of the valley to ridge land above.

There they built a beautiful farmhouse, a huge barn and other outbuildings, and a windmill to pump water. They also planted a large apple orchard. The valley and steep hills were used for cow pasture, and the rest was plowed and planted.

And it seemed to me that Grandma had always been cranky, for as long as I could remember.

A Time To Weep

"Leona, go see your Grandmother for awhile, you need to pay her a visit," Mother said, just as I was about to escape outside. We'd just arrived at Uncle Ernie's place a few minutes before Mother gave me my orders. I turned around and gave Mother a tired, pained look, like fulfilling this command was going to take a toll on my health.

"Oh, go on now," Mother scolded, ushering me down the hall. "I'll be in there soon as I can. Daddy and Uncle Ernie need to discuss some things with Aunt Meta and me for a while. So you go, now."

"Do I have a choice?" I asked, sarcastically. I knew I didn't, and reluctantly made my way down the hall toward Grandma's room. I lightly tapped my knuckles on the door, hoping she wouldn't hear me and I wouldn't have to go in. Then I could tell Mother that maybe Grandma was sleeping and I shouldn't disturb her just now.

No such luck.

"Who's there?" a scratchy voice sounded from behind the wooden door.

"It's just me, Leona," I said hesitantly, my brow creased with wrinkles of worry.

"Well, come in, child," Grandma said. I slowly opened the door, a rush of adrenaline filling my cheeks, and my heart starting to beat a little faster.

"Hello, Grandma," I said, standing in the doorway and noticing all the Catholic icons displayed in her room. Grandma was sitting on her bed.

"Come in, child, and close the door, for heaven's sake. Don't just stand there."

I closed the door quickly and walked over to a straight-backed chair in the corner of the room, sat down with my legs dangling over the edge, and tucked my hands under my legs. Grandma was a large woman, with long, gray hair that she braided and wrapped around her head. She was very stern and could be domineering at times.

Sitting in the chair, I watched as Grandma melted some wax over her kerosene lamp, which she applied to her leg sores to dry them up. Grandma had diabetes—the cause of the leg sores—and she certainly didn't feel well.

I could hardly blame her for being crabby and not easy to get along with, but still I had a hard time forcing myself to make conversation and act pleasant like Mother expected of me. I would rather have been doing other things.

"Where's yer chubby little sister?" she asked.

"Lenny's with Daddy and Uncle Ernie."

"What's she doin' with the men?"

"I don't know," I stammered.

"Stand up and let me take a closer look at ye."

I reluctantly stood up.

"Come closer. My eyesight ain't so good, you know."

I moved closer, knowing full well what was coming next.

"Hold in yer stomach child, it's stickin' out like a sow's belly."

I sucked my breath in hard, trying to draw my stomach into my ribs.

"There, that's better," she said. "Take those awful drawers off and try on these bloomers I'm makin' for ye," Grandma said, as she put down her wax pot and reached for the bloomers made of flour sack cloth, which were draped over the back of her chair.

I hesitated only a second, then slipped my overalls off. This was terribly embarrassing. I just hoped no one walked in on me like this. Grandma handed me the bloomers that still had pins in them and I pulled them up over my scrawny legs. The pins poked me a bit, here and there. I scratched my waist, then my leg and my arm. I couldn't help feeling itchy everywhere.

"Hold still, or I'll end up stickin' ye with a pin."

Suddenly emboldened, I said, "Grandma, it's the pins that's stickin' me now." I tried to hold still. I hated this. I just wanted to go outside and play.

"Suck in yer stomach, child."

Again, I sucked in my breath as hard as I could. Grandma stuck in some more pins around the waistline. I was still holding my breath when I started to feel faint. It was awful hot in Grandma's room. Suddenly, I couldn't hold it in any longer and I let my breath out all at once, with a gasp. My stomach popped out to its normal place, causing Grandma to drop her pins.

I heard the pins fall to the floor, tinkling on the hardwood boards during an excruciatingly long moment of silence. I squinted my eyes, hunched my shoulders, and wrapped my arms around my stomach, instinctively protecting my vital organs.

"Heaven's sakes, child, look what ye done now! Ye've got to learn to hold in yer stomach and act like a lady, or ye'll never find a man someday! Boys don't take to tomboys, ye know? They like feminine, pretty girls, who they can proudly take home to show their mothers. If ye keep going the way you are, child, I'll eat my words if ye ever find yourself a husband that'll put up with yer shenanigans. Only the good Lord knows what's to become of ye. Now for heaven's sake, hold in yer stomach so ye don't slouch, stand up straight and try to hold still for once."

Grandma tried again to pin the waistline, and this time I succeeded in holding my stomach in until she was done.

"All right now, take it off and I'll finish sewing this later. Right now, I gotta tend to my sores."

I quickly slipped off the bloomers and grabbed my comfortable overalls, gratefully sliding my legs into them with ease. Grandma hobbled back to her kitchen area and brought out a box of soda crackers and some butter. She buttered several crackers, put them on a plate and held it out so I could reach.

"Here, have a cracker. Looks like ye could use a little extra fat on yer bones. Goodness, you are brown. What are ye? Part Indian?"

I was kind of hungry, and gratefully took a couple crackers.

Grandma walked over to her radio and turned it on. The radio was pre-set to the station that aired "Ma Perkins." Grandma had to listen to it every day. She shuffled back to her bed, sat down, lifted one leg onto the bed and pulled her stocking down, revealing several large, red sores. Then she picked up the wooden stick she used to smear the wax on with, and began to cover one ugly sore at a time.

"Now go on, leave me alone so I can finish doing my legs," Grandma said, and I left immediately, thanking her for the crackers. I closed the door behind me, breathing a sigh of relief.

As I walked past the kitchen table, Daddy, Mother, Uncle Ernie and Aunt Meta were discussing something important.

"I think it would be best for all of you," Daddy said.

Uncle Ernie was agreeing.

"Yeah, she's a bit difficult to live with. You can't blame her for bein' this way. I'm sure I'd be just as ornery if I was in pain all the time. But this is awful hard on Meta and the girls. If we build Grandma her own little house, I think she'd be happier, and it'll be a relief to Meta, and yet Grandma will be close enough to holler if she needs help."

"When should we start?" Daddy asked.

"As soon as possible!" Aunt Meta said.

"Well, how about we start next week? This'll give me time to get the lumber together, timbers, nails," Daddy said.

"Ok, sounds good. I'll call Anton and see if he can help too," Uncle Ernie said.

So it was settled. Daddy and Uncle Ernie were going to build Grandma a little house of her own, out in the apple orchard.

I went outside to find Lenny and our cousins, Bernice and

Lenny (left) and Leona (right) flank their friend Betty Peterson, who lived in Boscobel and went to Sunday school with Leona. As one who lived "in town," visiting the Ahnen farm was Betty's only exposure to farm life. The older girls were both about twelve years old, and Lenny was seven when this shot was taken in one of the back pastures. They weren't really hunting, just posing with unloaded weapons for the camera, though Leona is holding her own .22 Remington rifle. Lenny and Betty hold two of Ralph's shotguns, and the dog is Drum, one of Leona's father's prized Walker Hounds. The animals lying in front of them are red foxes that Ralph had trapped for their fur. He hunted in the winter because the furs were thicker then and brought better money, which sometimes bought their groceries.

Beatrice. We were going to make the long trek down the hill to the old homestead in the valley below. Once we got down there, we'd be able to look up high, to a certain very steep hill, where there was a huge rock that was more like a cliff or an outcropping. On top of that rock, in the early spring, we could find what we called, "Easter flowers." They were really wild crocus—the first I'd ever seen—and I thought the yellow and purple blossoms on their fuzzy stems were gorgeous.

Sometime later, in mid-April, I came home from school one afternoon to find the farm deserted. Somehow, I knew they were all over at Grandma's house. Mother had told me that Grandma contracted pneumonia, but I'd kind of forgotten about it until this.

I quickly did the chores, including the milking, then ran all the way over to the home place. I saw the partially built house in the apple orchard as I trotted up to the farmhouse. I stopped at the door to catch my breath, then carefully walked in and on to Grandma's rooms.

I saw Mother and Lenny standing on one side of Grandma's bed, and Daddy sitting on the edge of it, his back to the door. Uncle Ernie and Aunt Meta, Bernice and Beatrice were on the other side. I walked up to Daddy and rested my hand on his shoulder. He looked up at me with sadness in his eyes.

Grandma was lying in her bed, her breathing heavy and labored. I didn't know if she even knew we were there.

I suddenly felt guilty and ashamed. I regretted all the mean thoughts I'd had about Grandma and wished I could take them back. I wanted to thank her for the bloomers, and for the soda crackers with butter that she gave me every time I visited.

But before I could say a word, Grandma drew a feeble breath, then let the air escape for the last time. I found myself holding my own breath, hoping Grandma would take another, but she never did.

Mother started to cry quietly, as did Aunt Meta and Lenny and Beatrice. Daddy and Uncle Ernie sat on the edge of Grandma's bed and bowed their heads. Daddy took Grandma's right hand and brought it to his lips. He kissed it tenderly, just holding her hand for a while, as he quietly prayed to our Heavenly Father to receive the soul of our dear Grandma.

Watching Daddy pray reminded me of a few weeks before, when we had gone to the Methodist Church in Boscobel for a revival. As usual, the revival was held at night, so the farmers could attend after their chores and milking were done. There was a traveling evangelistic speaker and some special music with different instruments. I was especially taken with the trombone.

The speaker was quite dignified, and nothing wild went on. After all, this was the Methodist church. But at the same time, he preached hellfire and brimstone, which visibly affected some folks. At times, he'd stop and lift his hands up to Heaven while he was praying for someone. He said, "Amen" and "Praise the Lord" so often, I lost count.

Towards the end, he gave an invitation for those who wanted to accept the Lord as their Savior to come forward. Daddy, Mother, Lenny and I were sitting toward the back, and we watched as men, women and children walked to the front of the church for the altar call.

Suddenly, Daddy got up and walked forward, too. He went to the altar to kneel in front of Reverend Taylor, and they prayed. Lenny went, too, and when she came back she said she had asked Jesus to come into her heart.

I wasn't sure what she meant by this, but I didn't ask any questions. I wasn't about to go up front, mainly because I thought I didn't have anything to repent for. I was bullheaded, and had the idea that becoming a Christian would curtail my activities.

After that, Daddy took his commitment very seriously. He had always smoked a pipe, but believed that if he was a Christian, this was not a good thing to do. So he threw his pipe and tobacco away and never smoked again. Now, just weeks after the revival, Grandma had passed away and Daddy was praying to his Savior in Heaven.

I wondered if Grandma really was in Heaven, her being so crabby and all, but I figured God knew why she was crabby and most likely pardoned her for this. Daddy finally stood up and said, quietly, that Grandma was with Jesus now, no more pain and sorrow, no more suffering.

After Grandma died, the undertakers took her body away, did the embalming, and brought her back in a casket to the house where she would stay for the next two days. Friends and family

This photo of a young Margie Wetter was probably taken around 1922. Leona believes her mother had it taken to give to Ralph as a gift, before they were married. Her hairstyle in the portrait reveals her German heritage.

Young Ralph Ahnen also had a special portrait made to give to his fiancée. These photos would have served much the same purpose as high school graduation photos do now. Ralph, like most of his peers, only attended school through the eighth grade, at the same Yellow School where his daughters would eventually go. The photo was taken in 1924, and he would have been 25 years old.

would come and go, bringing food, and there was always someone keeping watch over Grandma's body, day and night. Grandma had been Catholic, and the night before her funeral, the priest came out from the church in town to say The Rosary.

After the funeral, folks came back to the house again, and we'd eat more food. And there was so much food! I couldn't eat enough to fill a bird's tummy, yet I hardly cried a tear. It all seemed so strange to me.

Not long after Grandma's funeral, the first day of May rolled around in 1938, and turned out to be one of the best Sundays we ever had as a family. Mother made a picnic lunch and we all climbed into the car, Daddy and Mother sitting on the front seat, Lenny and I in the turtle. The car did not have windows, and it should have had side curtains, but these were missing. It was our car though; all ours, bought and paid for.

Daddy wanted to have our picnic at a special spot he knew about. He was very familiar with the land and woods that surrounded us, even our neighbor's property. Daddy seemed to have an innate sense of direction. You could set him down anywhere in the woods, blindfolded, and soon he'd know where he was; which way was south, where the creeks were, and how to get home. As a boy, he had roamed these hills and valleys, hunted in them, and had grown to know the land like the palm of his hand.

It was a warm spring day as we drove up our steep driveway and out onto the main road, heading toward Daddy's secret spot, watching for birds along the way. Most of the roadside fences were obscured with brush and vines, elderberry bushes, wild grapes, and some small trees. Bluebirds nested in hollow fence posts, orioles perched in the tall trees, and meadowlarks were everywhere.

At home, we had wrens in little birdhouses by the house, and they'd sing all day long. In cow pastures, the killdeer nested, and if you went too near their nests, the mother bird would pretend to be wounded, flopping along on the ground, leading us away from the nest. Of course, it worked every time. We'd follow her and ignore the nest.

Over the years, though, I found lots of killdeer nests, sometimes with a huge baby cowbird inside, instead of the expected occupants. Cowbirds did not make nests of their own, but laid their eggs in any nest they could find. After depositing its eggs, the cowbird would fly merrily on its way, leaving the unsuspecting killdeer parents to raise its young.

We always watched eagerly for the first sign of robins in the springtime. After a long, hard winter, Robin Redbreast was a welcome sight. Later in spring, we would hear the call of the whippoorwill. Daddy always said, "It's time to plant corn when the whippoorwills sing."

But today there was no planting corn for us, just a pleasant Sunday drive to a secret spot in the woods, where the violets would be blooming like a purple carpet underfoot, and hopefully the fish would be biting.

Daddy loved to fish, but the only chance a farmer had to go fishing was on rainy days. Sometimes he would take us to a nearby creek where we fished for suckers, but usually he would head for the Wisconsin River, about five miles away. There were times when we'd be coming home from a fishing trip with the sun shining and our neighbors working in the fields, after we had left in a downpour earlier that morning.

Daddy had no fancy equipment, just a cane pole with a cork and a hook, and any fish he caught tasted good to us. So, as Mother laid down a blanket under an old oak tree, Daddy, Lenny and I set about fishing in the creek. I dug around in the worm can for a juicy nightcrawler, pulled it out and laced it onto Daddy's hook. He threw the hook, nightcrawler, line, and cork out as far as he could into the water. Lenny and I sat on both sides of Daddy, watching eagerly for the cork to go under. We waited and waited.

"Maybe it's not the right time of day," I said after a few minutes of waiting.

I never did have much patience when it came to fishing. Daddy shrugged his shoulders and continued to watch with calm resignation.

"See that huge rock outcropping, way up yonder?" Daddy asked us, pointing towards a hill off in the distance. It was visible from a long ways away.

"Yeah, I see it. What about it?" I asked.

"Well, years ago when I was just a boy, I climbed to the top of that rock, called Easter Rock. And it was from Easter Rock that I saw my first automobile ever, driving right up Homer Road!"

"Really?" I asked.

"Oh, yeah. And I determined, right then and there, that someday I was going to buy myself a car just like it. It's taken a few more years than I anticipated, though." Daddy nodded his head and continued to wait patiently for a fish to nab the end of his line.

I took my shoes off and stuck my toes in the cold water.

"Want to hear another story?" Daddy asked.

"Yeah!" Lenny and I chimed in, together.

"Well, a long time ago, when I was a just a wee boy, I was supposed to be getting ready to go to church. This was before we had moved up on the hill. But me and my brothers, we sneaked off just before church to go hunting. It was then that I shot a wolf! Boy, was he a dandy!"

"You did?" I asked, incredulous.

Daddy nodded his head and fell silent again. Then something else occurred to him.

"Did I ever tell you why my handwriting was so bad?"

"No, tell us!"

"Well, when I was a boy in school, I was naturally left-handed, but the teacher forced me to write with my right hand. In those days, you did what you were told or you got your knuckles rapped with a stick! And I got my knuckles rapped plenty of times, because it was so much easier to use my left hand. I did learn to write with my right hand, but never very well. Couldn't!"

"Why'd she do that?" I asked, demanding an explanation for Daddy's sake. I felt so indignant at the thought of a teacher being so cruel to my Daddy.

He just shrugged his shoulders and fell silent again. After this, Daddy didn't offer any more stories, so I got up and began to wander.

Mother was arranging our picnic food in the middle of a blanket, humming to herself. She was laying out fried chicken and slices of bread with lots of butter on them, homemade pickles and spice cake.

I looked back at Lenny and Daddy. Lenny was leaning against Daddy, her head resting on his shoulder, just watching the water drift on by. Still no fish.

I followed along the banks of the creek, jumping it every once in awhile, when the two banks were close enough. I came upon a patch of spring violets, like a purple carpet. I picked a large bouquet for Mother.

Pretty soon I had wandered into what was known as Gunderson's Woods. It had been cut over decades ago, but was grown up again into a mature forest. Wild geraniums and wild columbine grew there, but the most beautiful flower I ever saw was an orchid called Yellow Lady's Slipper.

The slipper-shaped blossoms were bright yellow with slender, twisted, brown petals that looked like ties for the slipper. I became obsessed with hunting these delicate flowers, which grew in clumps of three to five, well hidden under the brush and ferns. I added one after another to Mother's bouquet. After half an hour or so of meandering, I finally ran back to find Mother and give her the flowers.

"Oh, these are beautiful!" she said, as she dipped her nose into the wildflowers. "Thank you, dear," Mother said, with tenderness in her eyes.

Boy, it felt good to be doing something like this with our folks! Lazy days down by the creek were a real treat. The years of hard work, the tiredness, the worries of the Depression all seemed to disappear for awhile, and we were carefree, happy and laughing, even if it was only for an afternoon.

Little did we know what was to come the very next day.

In happier days, a nattily dressed Ralph Ahnen poses in front of his old Model T Ford for his sister, who was playing photographer around 1922. This was probably his only good suit of clothes, and it's very likely he was on his way to court young Margie.

"Good breakfast, Mrs. Ahnen. I'd like to just sit here and enjoy my coffee with you gals, but I have to get outside. Say, did you hear the whippoorwills last night? You know what that means, don't ya?"

Daddy smiled at Mother. He looked at her teasingly and she started to blush. Mother quickly turned around and busied herself with some mindless task.

"It's time to plant the corn," Lenny and I chimed in together. Daddy laughed as he got up from the table and put on his boots, then went outside to the barn.

When The Whippoorwill Calls

"Here comes Masel, girls. Hurry up and finish your breakfast or you'll be late," Mother said.

Masel knocked on the screen door and then walked right in.

"Mornin', Mrs. Ahnen!" she said quickly and cheerfully.

"Good morning, Masel. The girls are almost done," Mother answered.

I swallowed my last bite of bread as Lenny was picking up her lunch and books, then got up to find my stuff, as well. The three of us headed out the door for school. Mother followed, then headed toward the barn.

"'Bye, Mother!" Lenny yelled, as we started to walk up the steep driveway.

I turned around to look for Daddy. He was over by the barn hitching up the horses, getting ready to pull the drag, or harrow, which was used for raking out newly plowed soil. Each section of the harrow was maybe five feet across, and Daddy had four of these, all connected together in a row. It took all four horses to pull the thing.

There were times when he'd just use one section and two horses, like if he was going to plow Janie's garden. But our fields required all the sections and all of our horses. This was hard work for him, as he would have to walk all day behind the drag in soft soil, and his knees had been hurting him lately.

"'Bye, Daddy!" I yelled, waving my hand. Daddy stopped to take notice and gave a hearty salute back to me, then turned to Cute

Ralph stands in the farmhouse yard, contentedly smoking his pipe, which indicates the photo was taken before his conversion to serious Christianity at a traveling revival. Behind him lies his Walker Hound, Drum, in whom Ralph took much pride and pleasure. The yard may seem a bit unkempt to modern, suburban readers used to well-groomed lawns. The Ahnens didn't own a lawn mower, so the grass was not regularly manicured, but that was common for this time.

to finish what he was doing. Cute was his favorite, and I understood why. She was a horse with personality and spunk, and would do anything Daddy asked of her. She was like putty in his hands. Her only downfall was that she could be a little skittish at times.

I stood still a moment, watching Daddy as he gently pulled the bridle up and over Cute's head. Then Mother walked up to Daddy, he turned around and they held each other in a brief embrace.

"What are ye waitin' for?" Masel asked.

I liked it when Mother and Daddy were affectionate with each other. It made me feel good. It didn't matter how poor we were, as long as we had each other, a roof over our heads, clothes on our backs, and enough food to eat. Just a man and a wife and their family, working the land by the sweat of the brow. Life doesn't get more basic than this, and I was satisfied with it.

"I'm coming," I said, breaking away from my folks and directing myself toward the uphill climb.

Twenty minutes later we were in school.

Ten-thirty was our morning recess time. We had recess for fifteen minutes in the morning and again in the afternoon, with an hour at lunchtime. This morning we were playing softball, and I was pitching. We were practicing for the big game at the end of the year, when all the parents would come on the last day of school for a picnic. We'd play ball and eat, and eat.

Daddy and Uncle Ernie would choose kids to play on their teams. I was especially proud of Daddy's ball playing ability, and naturally wanted him to pick me. One year at the picnic, Daddy chose my cousin Norman instead of me. It seemed like my heart would break, but I never told anyone. He certainly had no idea how I felt, and I was not angry, only felt again that being a girl had its disadvantages.

Lenny and Gen were swinging on the only two swings we had, which hung from a metal frame. The boys thought if they pumped hard enough they'd loop all the way around. Sometimes we'd play "Ante-I-Over," or hide and seek, but today was softball, and even Miss Fry was playing.

"Come on, Leona, put 'er right there," Norman said, indicating a spot in the air with his bat. I wound my arm several times just for show, then let the ball go overhand, with all the strength I could muster. Usually, I pitched underhand, but this time I really wanted to nail Norman. Unfortunately, my plan backfired.

The ball went directly where Norman had wanted it. He swung, and hit the ball with a crack! It flew through the air, way over our heads, sailing over the outfielders and well into the field. Kathleen ran as fast as she could to retrieve the ball, when suddenly we heard a car coming down the dusty road at a high rate of speed.

This kind of hurrying was unusual in the country, and everyone stopped to see who it was. I recognized the car right away. It was Uncle Ernie's, and he was heading for our road.

"What's up with Uncle Ernie?" I asked Bernice, who was standing at first base. She just shrugged her shoulders.

In the meantime, Norman took advantage of our distraction and made a home run, sliding into home base and destroying our home base cow pie. We were about to tell Norman to go look for a new one, when Miss Fry said it was time to go in.

Class was quiet that day, with students buried in their work, doing sums or memorizing spelling words.

"Mischievous. M, i, s, c, h..." I spelled the word out in my head as I looked around the room.

Norman was at the blackboard, writing down his arithmetic problems. Kathleen had her nose buried in a book. Lenny was up by the teacher's desk asking a question. Bernice and Beatrice were quizzing each other on their spelling words. Masel was writing fast and furiously in her notebook. Gus and Virgil were pretending to do schoolwork in the back row, but I knew better.

This was a free time, when we could work on anything we wanted, as long as we were being quiet and getting some work done. It also gave Miss Fry a chance to get some work done, herself.

Suddenly, we heard someone drive up in a car, park, and come running up the school steps. The door opened rather quickly. Uncle Ernie rounded the corner, looking harried and sweaty. We saw bloodstains on his shirt.

Miss Fry looked up from her desk, startled by Uncle Ernie's sudden entrance. He spoke first.

"Sorry to disturb your class, Miss Fry, but I need to take Leona and Lenice home right away."

Miss Fry looked concerned.

"Is everything okay?" she asked hesitantly.

Uncle Ernie shook his head, mumbling, "I… I really don't want to say right now, Ma'am."

"Oh, okay. Girls, don't worry about homework tonight. You can make it up another day."

Lenny and I put our books back in our desks and got up to go with Uncle Ernie. I looked into his eyes. They were red, filled with utter sorrow, and there were dirty tearstains on his cheeks.

Uncle Ernie avoided my gaze as he ushered us out the door and into his car. He started up the car and headed for our house without saying a word. I guessed if I was going to find out what was wrong, I was going to have to ask.

"Uncle Ernie…" I began to say, but my voice cracked and I stopped.

Uncle Ernie shook his head from side to side.

"I don't understand it girls. It's not fair. It's just not fair!" He slammed his fist down hard on the steering wheel.

Lenny and I both jumped in the seat. We'd never seen Uncle Ernie so tense and angry. What was going on here? What had happened? Was it Mother? Daddy? Had our house burned down?

If that was it, well, I could live with that. We'd build again, a newer, fancier house. Mother could then buy the new furniture she'd always wanted. That must be it! Our house must have burned down! Hopefully, Mother and Daddy had been outside when it happened.

But as we drove down our steep driveway, down into our farm, I could see immediately that the house was still there. The barn was still there. The horses were standing by the barn, still in their harnesses.

Funny, I'd thought Daddy was going to be out in the field all day. Oh no… Mother! It must be Mother. Why else would Daddy have come in from the field so early?

"Uncle Ernie, I have to know! What's happened to Mother?" I said, terribly frightened.

Uncle Ernie shook his head again, tears coming to his eyes.

"No, Leona, it's not your Mother, she's fine. It's... it's your Daddy."

Ralph and Margie's wedding portrait.

My heart stopped. I felt a terrible ache in my throat, and I couldn't breathe. Suddenly I felt sick.

Lenny was starting to cry. I put my arm around her, as I tried to ask questions of Uncle Ernie. But before I could say another word, the car stopped. We were home.

I opened the door, helping Lenny out and turning to take her hand. Feeling her small, pudgy hand in mine, I wanted to protect her from whatever we were about to experience, while feeling very small and scared myself.

Uncle Ernie quickly walked around the car and came up behind both of us. Resting a hand on each of our shoulders, he guided us toward the house. Just as we started to go through the door, I could hear Mother crying and I stopped dead in my tracks.

"I can't. I can't go in there."

Uncle Ernie walked around in front of us, bent down on one knee and looked me in the eye.

"Leona, something's happened to your Daddy. He's had a bad accident. His foot got caught in Cute's harness. It must have startled her and she bolted for the other end of the field. That damn horse has always been a bit skittish."

I just stared into Uncle Ernie's blue eyes, not truly seeing him, only hearing his words from a distance.

"Now, your Daddy's pretty banged up and he doesn't look like himself..."

But before Uncle Ernie could finish, I pulled away from his hands and ran for the parlor. I stopped in the doorway when I saw Daddy lying on the cot.

But it wasn't him, couldn't be him. I couldn't recognize him at all!

Mother got up from her knees and walked over to me. She took my hand and gently guided me over to the man lying on the cot. I stood stiffly by Mother's side as she whispered to the man, "Leona's here, darling."

The broken man moaned, trying to move his eyes, trying to find me. But one eye was swollen shut, the other bloodshot. I didn't think he could see me at all.

"Leona, he can't see you too well, but he can hear," Mother said, her eyes pleading with me.

Leona's Uncle Ernest was two years older than his brother, Ralph. Here, he holds Topsy (left), the horse that Leona rode from the time it was a year and a half old, and Lady, who came after Babe and Cute were gone. The picture was taken in the Ahnen's pasture, down by the water pump. In the background of this picture, on the left, you can see the lime quarry that Ralph had started with his other brother, Anton. Ernest was about forty years old in this picture, which means this photo was taken not long after he picked up Leona and Lenny from school the fateful day his brother died.

I knelt down on the wooden floor next to the cot. The man reached for my hand, and I took his. He squeezed my hand firmly. I was bravely trying to hold back the flood of tears that threatened to break over some inner barrier. I looked into the strange man's eye and finally saw Daddy; my brave, strong, loving, affectionate father, the man I followed everywhere, the hero of my world.

The life behind that one eye was ebbing slowly back to where it had come from.

I could tell Daddy recognized me, for he squeezed my hand once more. Then the rising flood overtook the levee, the tears coursing down my face, dripping onto Daddy's bloodstained hand.

Lenny knelt down next to me and buried her face in the blanket they put over Daddy. Mother moved to the other side of the cot and took Daddy's left hand in hers, kissing the back of his hand urgently.

I heard myself saying, "Daddy, I love you..." as he closed his eye, releasing his last breath into the air, and his soul into heaven.

Mother refused the first ambulance, Bartos. Somebody had called them by accident, and it wasn't the one she had requested. Finally, Fred Thoni arrived and they took Daddy away. By then, the Keltys and other neighbors had come to mourn with us, but I wasn't sure who was there and who wasn't, everything was such a blur.

I found myself, after awhile, wandering outside. The sun was starting to go down. Uncle Ernie and Uncle Anton were out by the barn, unharnessing the four horses. I walked up to them.

"What happened?" I asked directly, anger in my voice.

They both stopped to talk to me.

"Well, darlin', I guess your Daddy was taking the horses down to the water tank for a drink before starting work this morning. He must've tried to get on Cute's back to ride up the hill," Uncle Anton explained.

"You know, his knees have been bothering him so," Uncle Ernie added, "And how skittish Cute can be."

Uncle Anton continued.

"And his foot must've slipped as he tried to get on. He fell, but his one foot got caught in the harness somehow. This, or something else, must've startled Cute, and she dragged him the full length of the field, beyond Hickory Hill, through the ditch and back again, stopping under the big elm tree."

"When your Mother noticed that he wasn't coming back with the horses like he should've been, she went to the barnyard and from there she could see the horse stopped under the big elm," Uncle Ernie said, pointing to the far end of the field where the huge tree stood. "I guess she ran down there but couldn't help him, so she ran back to the house and called me."

"Is that why we saw you driving by so fast at recess time?" I asked in a cold and calculating voice, almost as if this was his fault. Uncle Ernie nodded.

"Yeah. I got your Daddy loose from the horse and we put him in my car and brought him back to the house." Uncle Ernie paused a minute to take a deep breath. "That's when I came for you and Lenny."

"At least we think this is what happened," Uncle Anton continued. "We can only guess, 'cause we weren't here when it happened, but from the look of things that's most likely..." He let the sentence drop, unfinished, as he turned away momentarily. When he turned to face me again, he was drying his eyes with a handkerchief.

I just nodded my head, understanding the whole story now. I turned to walk away, blindly, not really seeing or caring where I was going. The sky was beginning to show signs of darkness. Before I knew it, I found myself down at the far end of the field under the big elm tree.

I looked for signs of Daddy's struggle. The evidence was everywhere. Dark spots of blood littered the soil, hoof prints scattered everywhere, footprints all around, tire tracks from the car. I followed the drag marks all the way out past our clubhouse, down into the ditch, and back again. I walked back over to the old elm tree and sank to the ground, just staring at the spot where Cute had decided to stop after dragging Daddy for who knows how long.

I dropped my head into my hands, trying to block out the visions. I looked up again awhile later to a dark sky filled with stars. After an hour of sitting in numb silence, I finally got up and started to walk toward the house, vaguely aware that Mother might be worried about me, might need me.

Halfway home, my legs gave out and I fell to the ground, suddenly overcome with grief. I yelled at the night sky, "God! Where are you?" I felt nauseated, and my stomach hurt so bad. "Why? Answer me! Why?" I demanded and begged at the same time, pounding the earth with my fists as hard as I could. I was engulfed in sorrow like I had never felt in my entire life. I felt like I had been beaten. My face hurt. My chest ached. My eyes were swollen. My whole world had come crashing down around me, and I felt so alone, so lost without Daddy.

A couple days later, Edna, a lifelong friend of Mother's who lived in Stitzer, took Lenny and I to town and bought us each a new dress to wear to the funeral. I can't recall Lenny's dress, but mine was a brown print.

Next thing I knew, we were sitting in the front pew of the Methodist church. I heard the Eversen brothers singing "The Old Rugged Cross," and then Reverend Myron Taylor gave a funeral sermon. The text I remember clearly because unlike most sermons, it spoke to me and gave me hope.

"Do not let your hearts be troubled. Trust in God; trust also in me. In my Father's house are many rooms; if it were not so, I would have told you. I am going there to prepare a place for you. And if I go and prepare a place for you, I will come back and take you to be with me, that you also may be where I am.You know the way to the place where I am going."

And where Jesus was, Daddy was. That I knew for sure.

People came from all over to Daddy's funeral. The cars, wagons, horses and buggies were lined up for a long ways, trailing out of the cemetery and down the road. He was a well-respected man, highly thought of by everyone, especially the Keltys. The day Daddy died, their grief was just as painful as ours. They felt like they had lost one of their own family, and they mourned for him long after he was buried.

After Daddy died, Grandpa Wetter wanted Mother, Lenny and I to come live with him. As much as I loved the pine knob,

I certainly did not want to leave our home, and I'm sure Mother felt the same way.

The next few years were terribly hard. With a succession of hired men, we tried to keep the farm going. Despite the fact that the recession was easing up, there wasn't enough money to properly pay the hired men, and we expected too much. There was no way they could take Daddy's place.

Mother's grief was overwhelming at times. She would cry and cry, and I think we girls could not possibly understand her grief and feeling of helplessness. We missed Daddy, but our lives went on.

One evening soon after the accident, while doing chores alone out in the barn, I suddenly realized that this was the beginning of the rest of my life without Daddy. Strange thing was, nothing had changed. The birds were still singing. Chores still needed to be done. There were cows to be milked, and chickens and pigs to be fed. People's lives went on just as before. And yet, everything had changed.

I walked outside to gaze upward at the stars, wondering if Daddy could see me. Then, off in the distance, the call of a whippoorwill echoed through the night.

It was time to plant corn.

Lee Bennett sits with the author in Lee's backyard in Land O' Lakes, located in northern Wisconsin.

Epilogue

In 1943, Leona Ahnen became the first in her family to graduate from high school. Lee and her mother continued to run the farm another five years, but finally sold it in 1948 and moved into town. Lenny had left home and was out on her own by then. While still on the farm, Lee had developed an interest in flying. After moving into town, that interest began to blossom. She took flying lessons, soon earning her commercial license and instructor's rating. She gave flying lessons and ran the Boscobel airport until 1958. One of her flying students, Harris Bennett, found his unconventional teacher attractive, and they married in 1962.

"Lee" and Harris moved north to Land O' Lakes, Wisconsin. There, she ran the King's Gateway airport for about seven years while Harris did electrical work and gave seaplane rides. They also sold timber to the fledgling log home industry.

When Harris retired and Lee stopped running the airport, they decided to see more of the world. Between 1965 and 1976, Lee and Harris bought two retired Greyhound™ buses and converted them into recreational vehicles, which they drove on their travels to Alaska, Mexico, and the southern United States. They spent summer and fall in Land O' Lakes, and winters in Arizona. Lee and Harris also raised a succession of Brittany Spaniels until Harris died of cancer in the winter of 2000-2001.

About The Author

Kelly Simons was born in Amery, Wisconsin, in 1968. After a childhood on a farm in Rice Lake, Wisconsin, she moved with her family to Watersmeet, Michigan. There, Kelly graduated as valedictorian of her high school's senior class in 1986, and went on to attend the College of St. Scholastica in Duluth, Minnesota and the Western Iowa Tech Community College in Sioux City. She currently lives in Land O' Lakes, Wisconsin, with her husband and two children, where she runs Morning Sun Photography studio.

Kelly's writing has appeared in *Country Extra* and *Silent Sports* magazines. *Leona's Legacy* is her first book.